Digging Up Texas

A Guide to the Archaeology of the State

Robert Marcom

Digging Up Texas

A Guide to the Archaeology of the State

Robert Marcom

A Republic of Texas Press Book

TAYLOR TRADE PUBLISHING

Lanham • New York • Dallas • Boulder • Toronto • Plymouth, UK

A REPUBLIC OF TEXAS PRESS BOOK

Published by Taylor Trade Publishing
An imprint of The Rowman & Littlefield Publishing Group, Inc.
4501 Forbes Boulevard, Suite 200
Lanham, Maryland 20706

Estover Road
Plymouth PL6 7PY
United Kingdom

First Taylor Trade Publishing edition 2007

Distributed by National Book Network

Library of Congress Cataloging-in-Publication Data

Marcom, Robert F.
Digging up Texas : a guide to the archaeology of the state / Robert Marcom.
p. cm.
Includes bibliographical references (p.) and index.
ISBN-13: 978-1-55622-937-4 (pbk : alk. paper)

1. Texas—Antiquities. 2. Indians of North America—Texas—Antiquities. 3. Excavations (Archaeology)—Texas. I. Title.

F388 .M37 2002
976.4'01—dc21 2002010241

The paper used in this publication meets the minimum requirements of American National Standard for Information Sciences—Permanence of Paper for Printed Library Materials, ANSI/NISO Z39.48–1992.
Manufactured in the United States of America.

Contents

Foreword

One of the biggest problems facing archaeology today is that most people simply don't realize that archaeological projects are going on in their own communities, quite literally at times underfoot. This is pretty ironic given that we've witnessed a nearly exponential growth in the amount of fieldwork undertaken around the country (and in our own backyards) over the last thirty years.

A major part of this problem is that archaeologists mostly write books and articles geared towards the interests of their peers, or within the bounds of management reports. So unfortunately, writing for the public often takes a back seat, due to a lack of funding or the time to write for general audiences.

Digging Up Texas is a big step in the right direction. Robert Marcom covers the entire story of the peoples and cultures of Texas, from hunter-gatherers chasing bison into arroyos over 10,000 years ago, to the Caddo establishing villages across East Texas during the last millennium, and to the relatively recent immigration of pesky Europeans and Americans. Few books attempt to integrate the complete history of the state in one place—but seen together here—the past comes alive as a vivid and exciting story, 500 generations in the making.

As well, the author rightfully points to the vast knowledge that devoted amateur archaeologists have contributed to the discipline through volunteering both their time and expertise. As an example, the annual Texas Archeological Society (TAS) field school continues to serve as one of the premier archaeological institutions in the country, weaving professionals and amateurs together on exciting research projects. The author shows that anyone with a passing interest can join along in this pursuit of the past.

Because public education is critical to the survival and future success of archaeology, books like *Digging Up Texas* will continue to bring new friends and faces to the field, motivating future generations of Texans to study and preserve the heritage of all the former residents of their state. Marcom is to be commended for gathering this fascinating story from the dusty pages of archaeological reports, providing a book that the general public will surely enjoy.

Jason M. LaBelle
Ph.D. candidate, Department of Anthropology
Southern Methodist University
Dallas, Texas

Introduction

The tightly packed shell midden, a platform of shells and debris on which Native Americans had lived, was more than 2,000 years old. My wife, Ruth, and I were employed by a contract archaeology company to excavate the midden and learn as much about these people as possible. The U.S. Army Corps of Engineers would soon build levees that would flood the area, and we had little time in which to glean whatever we could from the remains of this ancient camp.

Ruth and I arrived with the rest of the crew on swamp boats that morning. She was part of the survey crew and busied herself with assembling the transit, the surveying rod, and other equipment. I helped unload the screens with which we would wash and sort everything that came out of the dig. When the surveying was complete, we would be able to decide the placement and dimensions of the "units." Then we could dig.

These prehistoric people had lived on the mainland, but they visited these islands. They collected clams and oysters, plentiful in the northern end of Galveston Bay. We might find shell tools, arrowheads, or personal decorations made from bone, shells, ceramic clay, and any other material available to a poor, hard-working tribe of Native Americans. Our work might add precious bits of knowledge about them and their lives.

Archaeology—the work is hard, the mosquitoes voracious, and the hours long. Only the excitement of discovery and a reverence for those human beings and their ways of life would suffice to enable us to bear the discomforts. But for us, those incentives are sufficient. I write this for those readers who would like to share the excitement. I dedicate it to the volunteers, known as avocational archaeologists, and to the professionals both of whom go home bone weary and intellectually rewarded.

Landowners in Texas deserve a special word of thanks, as well. Most archaeological sites in Texas are located on private land. The vast majority of cultural information gained from archaeology in the state of Texas comes through the generosity and understanding of the owners of the land on which sites are found. Myths abound among those landowners who are not educated in the laws and statutes of Texas. The myth, for instance, that land can somehow become public domain because it contains artifacts is one of many reasons sites go unreported. Archaeologists, and all the Texans who appreciate the archaeological heritage of our state, owe a debt to those owners who educate themselves and tolerate the eager intrusion on their property and privacy in order to expand knowledge of those who preceded us in Texas.

The purpose of this book is to inform and titillate: If you are one who has said you would like to learn something about archaeology, or even experience a dig, then this book is written for you. If you are interested in ancient Texas, historic or prehistoric, I hope to entice you with an incentive to learn more through study and participation. The emphasis here is on avocational archaeology (amateur association with professional archaeology) because that is the more common level of participation. You may find this book useful, as well, if you contemplate a professional career. The work described at the various sites detailed here is mostly the result of both professionals and avocationalists. Most of the professionals quoted herein readily acknowledge the avocational archaeologist as a valuable member of the archaeological community.

Chapter One

Past Lifeways, Texas Style

Archaeology! What magic the word has. It conjures images of daring adventurers hacking their way through dense jungle, seeking the Lost Jewel of Mumbo Jumbo. In truth, the science and craft of the archaeologist is not usually so dramatic as searching out a lost jewel or golden statue, but it can be every bit as interesting and far more significant.

The lands of Texas have a long history of occupation by a great variety of people. Those people brought with them their societies. They founded cultures here at least 13,000 years ago. The science of archaeology studies the artifacts of those societies and cultures, and of the succeeding cultures of people who made their homes and livelihoods in a given place.

Nothing could be more fascinating than the sweep and panorama offered by the lives of those first "Texans." Descended from paleolithic Asian people, they migrated from Asian steppes, across the Bering Straits; or they possibly arrived on the shores of North America after perilous voyages in canoes. They would have hopped from frigid island to island, and from beach to beach, for thousands of miles. Populations increased in those bountiful northern lands, forcing their children south. In time they came to settle on the fertile high plains around the Canadian River in the area now called the Texas Panhandle, hunting the great

Bison antiquus, and mammoths, evolving strategies for living prosperously in a new world.

Beginning with Chapter Two, the evidence of those paleolithic (old stone-age) people is examined. They hunted now-extinct species of animals for which human beings were a new and unexpected threat. Those early Texans found them a ready larder until the animals disappeared. Mammoth, mastodon, prehistoric horses, and other prey animals browsed the steppes and plains of Panhandle Texas, and until human beings came on the scene, they had only the saber-toothed cat and lesser predators to fear. According to the traditional interpretation, once they disappeared, the ancient Native Americans turned to other, smaller game. The atlatl (pronounced at-LAT-tull), a short spear thrown with a throwing stick, increased the range of their spears, and the paleolithic people learned to hunt the now wary game efficiently. They learned to domesticate crops and gathered plants and minerals, which they traded with other people for goods and materials they did not have. They evolved cultures that adapted to each new circumstance in a changing environment. Ice-age glaciers receded, leading to a warmer climate, and the cold steppes became warm, grassy plains. The plains cultures invented the bow and arrow, pottery and weaving, and were fully developed by the time the first escaped Spanish horses were captured.

Chapter Seven details the evidence for the impact of the early colonization by the Europeans. Both the French and Spanish made inroads into Texas and met Native Americans who made their homes in Texas. The explorers found them everywhere, from the East Texas woodlands, to the coastal plains, bays, and beaches, to the far western deserts and the Texas hill country. Every kind of Texas countryside had its complement of Native Americans, each with their own society and survival strategies.

To this, the Spanish, French, English, and later, the *Norte Americanos*, added their own lifeways. European-style villages, towns, and cities intruded on and co-opted older Native American villages. Farms and ranches replaced the open plains, forests, and fields. Domestic cattle and horses replaced the gigantic buffalo herds. Each society of humans left its own artifacts and evidence behind. The delightful job of the archaeologist is to discover as much as possible about these people—their epic struggles and their determined persistence.

≋≋≋

Quite apart from the cocktail party version of archaeology, the real thing is done in the dirt, outside, and in all kinds of weather. It almost never involves quantities of precious metals or jewels, and it always destroys the object of its investigation: the site that is excavated. Archaeological excavations are ideally done by professional people with a commitment to scientific principles. The archaeology professional has a university-level education that includes academic instruction and field experience. The archaeologist in charge will usually have advanced degrees and considerable experience in one or more specialty areas of the discipline.

Archaeology is the study of artifacts, traces, and material remains in order to understand lives and cultures of people. The artifacts of the many peoples of Texas include practically the entire tool and implement assemblage of ancient and historic human beings. Simple cobble tools and digging sticks, napped chert (the same material as flint)—from the crudest forms to some of the most elegant in the world—copper and turquoise jewelry, finely decorated clothing, and most every variety of firearms, European precious metals, coins, ships, furniture...the list is nearly endless.

Different philosophies approach archaeology in different ways, and we will discuss some of these approaches in detail later in this book. It may be helpful to preview the different approaches here in order to provide some context. (Note: the spelling of archeology versus archaeology will appear interchangeably in this book. Both are correct and the spelling may change according to the different persons quoted, or according to the official titles of organizations cited.)

Classical Archaeology

This is the approach with which most people are familiar. The earliest classical archaeologists were confronted with ancient monumental remains for which they had no history. Pyramids in Egypt intrigued the military forces of Napoleon Bonaparte in the eighteenth century. Ruins of giant buildings and step-pyramids for which the Spanish *conquistadores* could see no explanation dotted the conquered lands.

The monumental architecture is usually associated with societies that ended hundreds or even thousands of years before their discovery. The great buildings and monuments were often built by monarchs, pharaohs, or religious leaders for their individual purposes, now obscured by passage of time.

The purpose of classical archaeology is to take the information gleaned from these remains of long extinct cultures and societies and "paint in" the missing historical past. The archaeologist may use historical documents, inscriptions, and historical concepts in order to frame an investigation. Archaeology is seen, in this view, as an adjunct to the field of ancient history. This view of the study of archaeology is commonly taught in European colleges and universities. The Mayan and Aztecan ruins found in Central America have often been excavated utilizing classical techniques.

Anthropological Archaeology

In this philosophical understanding, artifacts and remains are studied in the context of culture. Societies of the past are analyzed through the cultural patterns that are demonstrated by inscriptions, writings, architecture, and physical remains. Areas of study may include linguistics, physical anthropology, statistics, and probability, as well as the traditional areas of artifact analysis.

The primary purpose of anthropology is the comparative study of human cultures and societies. The common method for reporting on these differences is the ethnography: a formal academic system of observations that account for difference in methods each culture uses to accomplish the tasks and duties all people have in common—eating and procuring food, dealing with dangers to the social group, living arrangements, mating and raising children, growing old, infirmity, disease, warfare, and leadership, to name a few.

Anthropological archaeology begins from an assumption that people, in the past, used similar methods for solving problems and creating social structure as do contemporary cultures. The anthropological archaeologist forms hypotheses and null hypotheses based on anthropologically valid concepts.

Historical Archaeology

Despite the implications of its title, this field of archaeology is actually a subset of anthropological archaeology. Historical archaeologists study the artifacts and societal remains of record-keeping cultures. Findings may then be analyzed in order to illuminate the facts underlying interpretations of historical cultures.

Biblical Archaeology

This field of study borrows from both classical and historical areas of study. The context of archaeological inquiry is usually text from the Judeo-Christian Holy Bible. The investigations may be begun with the intention of questioning, or with the idea of confirming the biblical text. In that sense, it is rather like classical archaeology. The artifacts and remains may also be studied in an investigation of cultural and societal relationships and resources. The result will be interpreted to better understand the lifeways of the population. In that sense, it would be similar to historical archaeology.

"Indiana Jones" Archaeology

This is the romantic view of archaeology engendered and perpetuated by popular fiction. In this type of archaeology, the hero/heroine braves exotic dangers, cheating death at every turn, in order to find fabulously valuable artifacts before the bad guys/gals can do so. The result of this adventure is often wealth, fame, and marital union with the co-star.

Professional archaeologists have a name for Indiana Jones archaeologists. They are termed "pot hunters." By pillaging artifacts, the Joneses destroy any value those artifacts may have in terms of contributing to scientific knowledge and understanding. Local governments pass laws against this sort of enterprise. Governments often have another name for the Joneses. They may call them robbers, plunderers, despoilers, or thieves.

Avocational Archaeology

Those who love the past and seek to understand the human heritage are often drawn to the subject of archaeology. Many are content with reading about it and viewing exhibits or interpretive sites. Others need to participate; they seek out archaeological associations in their area and join them. They take advantage of opportunities to participate in digs and to learn the processes and techniques of archaeological study.

The avocational archaeologist may come to the field with no experience or formal education, but they often gain both. Archaeological associations offer education through seminars, speakers at their meetings, and information on local college classes. Experience may be gained through active participation in excavations and laboratory field schools.

What avocational archaeology is not is artifact hunting, self-supervised excavation, arrowhead collecting, or purchasing artifacts on the open market. None of these activities further humanity's knowledge of past lives and human endeavor. All of these activities are used for personal acquisition at the expense of understanding our archaeological heritage.

While it can be quite fulfilling and personally gratifying to find a beautifully flaked biface and to show it off to your friends, or to display the frame of an old pistol recovered from a battle site, the information these wonderful artifacts might have yielded to trained investigators is lost forever. Avocational archaeologists understand that the personal satisfaction gained is not worth the loss to human knowledge.

Principles and Process

The purpose of this book is to survey the archaeological sites of Texas from an avocational perspective and learn of the many peoples and their lifeways: their social, cultural, behavioral, and occupational activities. It is not the purpose to teach archaeology here. There are many who are more qualified and capable of imparting a fine education in the science and techniques that an archaeologist requires for plying the profession. Instead, I hope to share with the layperson the excitement and wonder inspired by the tools, dwellings, personal effects, and trash left behind by other people, in other times.

While it is beyond the scope of this book to attempt a complete understanding of the underpinnings of archaeological science, a certain familiarity with the underlying principles will help the reader to understand some of the processes and conclusions discussed here. Archaeologists make certain assumptions, sometimes called first principles, upon which they build. They are often simple in their conception and may even seem completely obvious, until one remembers that these principles have not always been in existence. If the discussion of these principles and the resulting tools and techniques are not of interest, this would be the place to skip ahead to the next chapter.

Principle of Superposition

When materials are deposited, the law of gravity will dictate that the younger material lands on top of the older material. This leads to layers of materials, called *strata*. The layers are interpreted as the *stratigraphy* (arrangement by layers) of the site.

My finest moment of epiphany for this principle occurred during a "dig" (archaeological excavation) at the old Levi Jordan plantation site. The plantation offered a unique opportunity for archaeologist and professor Dr. Kenneth Brown to design a scientific plan for the excavation of a group of slaves' quarters and outbuildings that had never been disturbed. According to the common understanding, the ex-slaves, now freed by the Civil War, were driven from the site without any of their belongings except those they could carry.

My appreciation of superposition came when I excavated a unit (a measured and surveyed square to be excavated) that contained, in progressive layers:

- Hand-made fired brick from the estate kilns
- Chicken bones
- Fork and spoon
- Tin plate
- Nails and unrecognizable iron strapping
- The debris pattern formed by the pattern of the room's floorboards.

Principle of Uniformity

This idea, borrowed from geology, is that processes of the ancient past are consistent with or "uniform" with processes of the present time. Some processes are so basic to human needs that they must contain uniformity.

Perhaps my most noteworthy moment of appreciation in regard to cultural uniformity came when I toured a dig on the Mediterranean Sea. The site was the Roman city of Caesarea Maritima in Israel. The first impression I gleaned from the arrangement of shops, restaurants, and homes was that the feel was similar to that of any nearby village.

Cultural uniformity implies that later cultures that are descended from, and related to, earlier cultures will conserve elements of the culture of their ancestors. Inheritance is a special relationship among human societies. We have some excellent models of cultural uniformity in the recorded histories of the dynasties of ancient Egypt and Rome, as well as in more modern cultures such as those of Western Europeans. Attitudes, implements, social values, and relationships may be uniquely characteristic of each one and different from most of the others.

The Scientific Principle

The scientific process is based in a principle that assumes that all processes in nature can be discovered, and that those processes can be reproduced. This principle is a foundational concept for anthropological archaeology. Much of the science for an investigation takes place in the laboratory.

Archaeologists may supplement, confirm, and discover evidence of relationships and technologies by use of scientific laboratory investigation. Often pottery fragments spread over many layers of deposit can be linked by chemical analysis to a single source of clay.

Artifacts may be subjected to chemical analysis in order to date origin and use. Radiological carbon dating, dendrochronology, type analyses, description, and cataloging are all done in the archaeological lab. The patterns of distribution and context will be carefully analyzed and combined with the laboratory evidence to enable an interpretation. This careful investigation of the evidence is framed by the archaeologist in her/his original hypothesis: a statement of what the site is expected to reveal.

A hypothesis is always balanced by at least one statement that, if true, disproves the hypothesis. This opposing idea is called the null hypothesis. No conclusion is accepted without a sincere effort having been made to disprove it.

≋≋≋

The well-prepared archaeologist needs skills and general knowledge of a wide variety of topics in order to understand the facts that artifacts represent. Knowledge from the disciplines of chemistry, geology, biology, botany, physics, anthropology, sociology, history, and other science specialties could prove to be essential to understanding the artifacts, their distribution, and their context. When detailed knowledge of a given science or art is required, archaeologists will call in expert consultants in the field. The expertise employed will often depend on the orientation of the "school" of archaeology to which the archaeologist subscribes.

There are two major schools or disciplines of the study of archaeology in the world today: the historical school, and the anthropological school. The historical school is primarily taught to and utilized by European universities, and the anthropological school is the one common to American universities.

The approach to the study of past lifeways in Texas and in the United States is from the perspective of the science and disciplines of anthropology. The study of the science of human adaptation and behavior has much to offer in the investigation of artifacts and their distribution and association.

Whether approached as a topic in the humanities (history, arts) or as a science (anthropology, sociology), the study of archaeology has a great deal to offer toward understanding who we are and how we came to be as we are now. The field of archaeology has offered answers for such intriguing questions as:

Who were the first Texans? Why did they come to the region?

When was Texas first occupied, and what did they want here?

Were the extinctions of the great Ice Age animals due in part to hunting?

What is the truth of the occupations by Europeans and their encounters with Native Americans?

What happened to the Native American people who lived in Texas? Where did they go?

But the questions need not be so far ranging or grand. For instance, there are university-sponsored projects looking at and cataloguing contemporary trash, in order to find out what may be understood about contemporary culture through examining its refuse. Clearly, archaeology is a science that can be used to increase human knowledge on many levels.

Processing Artifacts

Archaeological information is gathered in a destructive process. Archaeologists are often heard to say that through the act of excavating a site, it is being destroyed; the information that is not recorded is lost forever. This sobering fact is kept firmly in mind by the archaeologist in charge, often called the principal investigator, or PI. No matter whether the site is an ancient trash mound, called a midden, a campsite, a butchering ground, or a historical building or battleground, the excavation must be carefully planned and its progress and results thoroughly documented.

Exact location, called provenience—associations between artifacts in vertical and horizontal space—and the conditions under which the artifact is found are carefully recorded in the field. These field notes are often supplemented with maps, drawings, and photos. If the artifacts are removed from the ground, they are placed in bags. Then the bag is carefully identified with the site name, usually a trinomial identification, and the provenience information: the unit number, the exact depth and level of excavation, and the horizontal location. Of course, if the item is not particularly unique, as with individual shells in a shell midden, then they might only be identified generally. Other artifacts are unique, such as projectile points or pocketknives, and will receive the full documentation.

The documentation does not end in the field. Once the artifacts are bagged, they are usually taken to the laboratory. Once there, they will be cleaned, stabilized when

Left: the water screen, used to wash the heavy clay through quarter-inch hardware cloth and find artifacts. Foreground: the gasoline-powered pump that delivered the saltwater from the bay to the screen. Background: one of two airboats that delivered the crew to the shell midden sites.

These artifacts have been washed and once dry, will be ready to be cataloged. The tray includes freshwater shell and chert debitage. Photo taken at the Texas Archeological Society 2002 Field School. Photos by Robert Marcom, 2002.

necessary, and identified. The cardinal rule for identification is if you're not sure, don't name it. Often the records of a given site may be used in analysis without reexamining the artifacts. If they are wrongly identified, the analysis will be based on faulty information. Rather than to identify a point as a Perdiz, or Alba, or even as broadly as an arrowhead, unless you are an expert it is probably more in the interest of accuracy to call it a projectile point and let the primary investigator draw her or his own conclusions.

Once the artifacts have been identified, they can be cataloged. In the current era, with computers available to practically everyone, the catalogs for sites can be made available quickly and easily. Archaeology has benefited in this and many other ways from the computer revolution. Artifacts, catalogs, maps, and drawings are often made available to qualified researchers at universities and labs that are specified as archives. Texas is fortunate to have several university archives, including the Texas Archeological Research Laboratory, at the University of Texas at Austin.

If the excavation has been properly documented and the information and artifacts accurately processed, and if the records and materials are carefully archived, then the data produced will be preserved through time. Though the site has been destroyed, the information gained will serve science throughout time.

Chapter Two

The First Texans

The paleoarchaeology of Texas begins with the first human beings to set foot in the area that would later become the state, and ends with the time when the first historical records are created. The term *paleo*, as it is used in this context, means "ancient."

Twenty thousand years ago, Texas, and the world, was a very different place. The climate was colder and wetter than it is now. Deciduous forests, full of cold-adapted hardwood trees, dominated much of the state. The sea level was lower, and the shore much further out toward the Gulf of Mexico than today's coastline. In the west and north of current-day Texas, tall, cold-adapted grasses were interspersed with woodland copses, making an ideal habitat for browsing herbivores, such as mastodon and giant sloth. The prairies were well watered, and grazing herbivores, like *Bison antiquus*, mammoth, and early horses had little trouble finding sustenance.

Paleolithic-age Native Americans, more commonly called Paleo-Indians (also written Paleoindians), arrived in the vicinity of Texas on foot, carrying all their worldly goods. These people made whatever they needed from the plants, minerals, and animals they found in their new homes. They were few and scattered, and so is the evidence of them that remains today.

The story of the first Texans begins not in Texas but rather on the other side of the world, in eastern Siberia. A cave excavated by Yuri Mochanov (Mochanov, 1977) gives evidence of the ancestors of the people who first crossed the Bering Straits. Those hardy big-game hunters expanded rapidly, populating eastern Asia during the last part of the Würm glaciation, about 20,000 years before present (BP).

Evidence that these were the first people to arrive on western shores includes such diverse factors including skeletal and dental similarities and virtually identical tool kits from the Dukhtai cave in eastern Asia, excavated by Mochanov, and tools found at Bluefish Cave in the Yukon Territory. The journey from Asia, across the land bridge of Beringia, and across Alaska to the Yukon must have been a long and arduous one. We can imagine what it may have been like for these hardy men, women, and children.

How it Might Have Been (fictional)

The desolation of the eastern lands was frightening to you, at first. Following the large herd of mammoths, your band of twenty-eight men, women, and children worked slowly, day by day, toward the pale, cold, rising sun.

"How much is left?"

You look up to see the patriarch and leader, Vla, looking down on you with concern etched in his weathered face. You see the grease, made from fat and charcoal and applied to prevent wind and sunburn, has stiffened his face, but the expression is unmistakable. For a moment, the sense of his question eludes you.

You are the Stone Spirit Listener. You finally grasp the meaning of Vla's question. Looking down to the symmetrical chunk of flint in your hand, you feel a sense of your value to the small band of hunters, and this family.

"We have this one and three more."

"Good. Better than I thought. Have you seen any sign of more flint?"

Front to back are Paleo-Native American, dire wolf (left), saber toothed cat (right), *Bison antiquus*, and Colombian mammoth. Remains of these animals, as well as extinct species of horse, have been found in association with human cultural activities. Drawing by Jason Eckhardt.

"No. Even the river has no stones. Only mud." You swat at a mosquito, larger than you'd known before this trek. "Blades can come from this core, but not points. There are many points left in these other flint stones."

"We will move tomorrow."

Vla seems confident to you. You are not as worried as you were at first. Vla has been very careful, and no one has died for a long time now. You turn the flint core, attuning your eyes and hands to its features. You began your relationship with the spirit in this stone when you first knocked off one end with the strong, heavy, black hammer-stone, making the flat platform. The spirit gave the stone to you: the platform was perfect. You then held the flint with the flat end up, striking the edge, turning the stone, striking again. With long practice and reverence for the spirit of flint, you removed the white outer coating of lime, revealing the glassy brown of the flint. Now, after many strikes of the hammer stone, you have a small pile of blades lying at your feet. Each one is about an inch in length and about a half inch wide. The parallel edges of the rectangular shape are extremely sharp. You pick up each blade; look at it in cross-section. They are thicker in the middle, tapering to nothing at the edge. You judge the blades to be good, one at a time.

You know the flint blades are essential to your band. They can cut meat and hide. They can scrape fat and sinew. In the hands of the skilled women of your band, they make clothes and they make meals. They make shelters and they make natural wood and bones into implements. They are as important—maybe *more* important—than the spear points and scrapers you make.

The next morning, at first light, your band is packed and ready to continue the trek. The mammoth herd is moving east, following the sluggish brown river and the low arctic grass that grows beside it. Away from the river, the ground is rocky and barren, save for sparse sedge and mosses. Nothing is important to you, though, except the mammoths.

The Medicine Woman, Vla's wife, stops occasionally. She collects small plants and other things you can't identify. She makes medicines that heal and cure. She talks to other spirits, spirits you can't see. Whenever she stops, Vla signals a young man to stay with her, while the other men move on, behind the mammoth herd. There won't be a hunt today. The band has plenty of meat, and plenty of vegetation from the stomachs of mammoth, and the group can only carry so much. When the herd moves, the band must follow. The mammoths are a walking commissary for your family group, and to lose contact with them would be a disaster.

Occasionally you range over to the riverbank, as the group moves along. You look for nodules that might contain flint. Though you don't see any, you aren't worried, yet. Not yet.

This detail of an exhibit pictures the bones of the extinct species *Bison antiquus*. The bones were discovered during the excavation of a Paleoindian bison kill site at Plainview, Texas. The bones are more than 10,000 years old. Photograph by Robert Marcom, courtesy of The University of Texas at Austin, Texas Archeological Research Laboratory.

This reproduction of a Paleolithic projectile point is placed in the position where an authentic point was found during an excavation. Photograph by Robert Marcom 2002, courtesy of The University of Texas at Austin, Texas Archeological Research Laboratory.

The Clovis People
(sites: Alibates, TX; Clovis, NM)

Archaeologists speak of particular assemblages of artifacts from a single culture as a *horizon*. The Clovis horizon generally consists of flint tools and skeletal remains of the butchered carcasses of extinct mammoths, mastodons, and bison. The fact that the Clovis people made a sudden, profound, if somewhat brief impact on North America is obvious in the archaeological record. But where did these accomplished hunters come from, and what happened to end their lifestyle?

Theories for alternative methods of travel to the Americas have been proposed; some theories are readily dismissed. For instance some have claimed that the great sculpted heads found on the coast of central Mexico represent a cultural transfer from Africa to Central America. The inference is drawn from the appearance and features found on the carved heads. The main problem for those who hold to this theory is the age of the heads. They are dated to around 2,000 years BP, and there is no other evidence for Africans in Central America in earlier times.

Other evidence may be forthcoming from "pre-Clovis" sites, where radiocarbon dating indicates occupation before the traditional beginning of the period archaeologists date for the Clovis people. The most renowned of the sites, the Meadowcroft Rock Shelter, in Pennsylvania, has yielded dates as early as 20,000 years BP. Here in Texas, the Gault site (which is covered in this book in detail) has materials dated much earlier than the traditional beginning of the Clovis culture. Some of the oldest dates of the Meadowcroft materials have been questioned because of the possibility of contamination with carbon much older than that, which resulted from human activity.

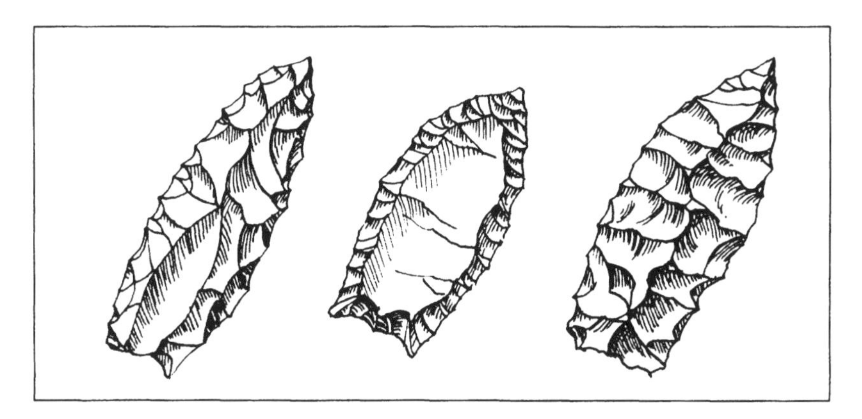

Paleolithic projectile points. Left: the Clovis point. Center: the Folsom point. Right: the Plainview point. All three of these classic forms are designed to kill big game. They require a high degree of skill to create, but the resulting amount of meat from megafauna prey more than justified the investment of time and effort. Drawing by Jason Eckhardt.

Clovis blades and tools made on blades. Photograph by Milton Bell. Courtesy of the Texas Archeological Research Laboratory, The University of Texas at Austin.

Such contamination would make a sample appear to date older than the activity that produced it. If, for instance, coal dust from a nearby deposit in Pennsylvania blew onto campfire coals of the Meadowcroft Rock Shelter, or was transported by groundwater through the deposits, the resulting dates could appear to be earlier than their actual deposit.

However, the evidence for at least one alternative theory is intriguing. There is at least circumstantial support for the theory that paleolithic people arrived on the eastern shores of North America as long ago as 15,000 years BP. Such an arrival would have required a sea voyage made during the Wisconsin period (same as the Würm glaciation period in Europe) of glaciation, between 25,000 and 12,000 years BP. Surprisingly, there is some support for ocean travel in this early period of time.

Evidence of the population of islands in the Pacific Ocean in Pleistocene times (before 11,000 years BP) include the peopling of Australia and Japan after 60,000 years BP. During this time, no land bridges connected these bodies with Asia. The simple conclusion is that people traveled by boat or raft to arrive at these places. The conclusion is made more astounding when one takes into account that hundreds of people must have made the trip, because a breeding population requires a group of men and women large enough to include enough genes to avoid extinction from inbreeding.

There is evidence of a curious distribution pattern for Clovis period stone artifacts, called *lithics* by archaeologists, on the North American continent. If the continent was populated from the direction of the Bering land bridge, one would expect to see the heaviest concentration of Clovis lithics radiating from that direction. This is not the case: The heaviest concentration of Clovis artifacts is on the East

Coast of the United States, and lessens as one looks to the west and northwest.

There is a significant caveat to offer here: The apparent distribution may not be the actual distribution. It may be that more Clovis artifacts have been found in the eastern United States because more archaeological work has been done in the East. The gradient of concentration appears to be too great for this to be the explanation.

The European subcontinent experienced a depopulation during the Solutrean horizon (Würm/Wisconsin glaciation), approximately 18,000 years BP due to an increase in glaciation. These people had a highly developed flint technology that looks very much like the early paleolithic points used by the earliest people in North America. Some archaeologists have called the two kits "virtually indistinguishable" one from the other. There is a caveat to be offered here, as well: when technologies are invented to solve a particular problem, form often follows function. It is possible that the two lithic traditions are similar because they were designed for the same purpose and from the same materials. Still, the confluence of lithic shapes in such a close time frame, and given distribution and climatic circumstances, makes for a strong suspicion of transplanted culture.

Beringia Land Bridge

While paleolithic people may have arrived in North America via other routes, it is clear that the bulk of the first people came across the Bering Sea land bridge, known as Beringia. The people who followed the Bering land bridge between Asia and North America may have been bold adventurers, opening new lands, or they may have had pragmatic reasons to take such enormous risks by leaving the familiar and striking out for unknown climes. The previous story takes the latter view. It is a fact that big game hunters must follow their quarry. These people in the tale inadvertently trekked to new lands; they both followed and preceded many such small bands. They probably encountered each other rarely, because human population was sparse and scattered between 20,000 and 12,000 years BP.

When they did come across each other, it seems likely they traded information and unspoken-for women. Some evidence for this might be found in the hospitality and marriage customs of the *Enuit* and *Eskimo* bands that populate the far north in current times. At some time after 12,000 years BP, the people who found their way onto the continent of North America must have spread to the south. The people known from their material remains as the Clovis Tradition were certainly among the most successful of those descendants.

The big game hunters came across the flat, featureless land bridge that was once the bottom of the sea. They found hunting grounds and places to live. The glaciers of the ice age known as the Würm in the lands they'd left and as the Wisconsin in the new lands, were not solid across the northern reaches of the areas of the Yukon and British Columbia. Their cooking hearths and mammoth game kills attest to the widespread habitation of the new lands, from Meadowcroft, Pennsylvania, to Central America before 11,500 years BP. That the Clovis spear point is found in these far-flung sites is symbolic of human ingenuity and determination. These capable sojourners created solutions for the worst of the problems they encountered in a strange land. They prospered and expanded, with their material culture evidencing little change in the archaeological record over a period of 500 years. To lend a bit of perspective, think about the numerous changes in Texas's material culture that have occurred over the *last* five hundred years.

We derive the name for the Clovis people from the most easily recognized artifact of their culture: the Clovis spear point. The finely crafted, symmetrical dart point was probably affixed to a short spear and thrown with a stick, called an atlatl. The point may have been used on a longer spear, rather than by use of the atlatl, a detail of method yet to be resolved by archaeologists.

The Clovis people hunted mammoths and were widespread. Clovis points have been found at sites from northern Canada to Central America and found at many of these sites in association with mammoth remains. There is little doubt that some, if not all, of the mammoths were killed by the Clovis hunters. The points have even been found lodged in mammoth bones.

Mammoth skull, drawn from a skull located at the Museum of the Llano Estacado, Plainview, Texas. Drawing by Jason Eckhardt.

The Clovis culture is documented by carbon-14 dating to have lasted for about 500 years, from about 11,500 BP to 11,000 BP. The first Clovis site was found just across the border, south of the town of Clovis, New Mexico. These people of the Clovis Tradition may well have been the very first people to live in the area that would become Texas. Imagine, then, what it might have been like to hunt these powerful animals.

How it Might Have Been (fictional)

You and your companion crouch, two hunters smeared with pitch and oils, two spears held ready. Four other hunters circle. The grass is tall and the air is clear. Eight mammoths graze peacefully upwind. Taller than two men high, the enormous animals seem imperturbable. The elephantine creatures move as a group, pausing every step to pluck a thick bunch of grass from the ground, shake the dirt from the roots, then stuff the vegetation into gaping mouths.

Dressed in a warm bison-hide tunic, lined with soft rabbit furs and shod with tough rhinoceros hide boots, you tense in anticipation.

"The spirits of the mammoth have favored our hunt," you say, with great gravity.

"May it be so," utters your young companion.

You intone, "Spirit of the Spear, make my arm true... Spirit of Flint, make my tooth sharp!"

"May it be so," the youth repeats, automatically.

As the herd turns, the other men will leap out, waving and shouting. If all goes well, the half-circle of hunters will close in and there will be only one choice for the prey: They will be driven toward the waiting spears.

As you look to the younger man for speed and agility, the youth looks to you, the older man, for knowledge. It has ever been this way among the tribe.

"Watch the old bull. You will see him warn the herd," you instruct.

"Yi-eeeie!" cry the chasers, and the rush begins. The bull trumpets. Instantly, the mammoths surge into a strangely graceful lope, the circle of men leaping after them through the grass. Six females and one old bull try to protect a calf. The calf is taller than a man. While the herd's wily matriarch shepherds the group, the largest bull turns suddenly and charges the onrushing men, its prehensile trunk held high and its tree-trunk thick feet gouging the sod. One hunter is too close; he flies through the air as the bull catches and tosses him with its enormous tusks.

"Run for the calf!" you shout to your young companion. "Look! The bull abandons the calf and drives the cows! They will run to the bog...we will kill this calf...." Your breath is short now, your legs pump. You are falling behind the younger man, despite all the strength you can muster. But the gods are with you; you see the calf turn suddenly, away from the young hunter and toward you.

You know this might be your last hunt, your last kill. A smile plays fleetingly across your reddened, wind-burned face; you're grateful to all the spirits for this chance. You heft the five-foot-long spear shaft, holding it lightly in its throwing stick, readying the razor-sharp four-inch flint point as you charge. Muscles tense as you prepare to strike deep through the tough hide of the immature mammoth. Running toward the mammoth from opposite sides, both of you adjust your stride. You must sidestep the oncoming animal at a precise moment then stab into its side as it rushes past. In past hunts many men of the tribe lost their lives performing this maneuver.

You feel the air burning in your lungs and the grass tearing at your legs. You feint to one side and the animal turns. Spinning, vaulting, you land with one leg stiff, digging into the loam. Uncoiling like a spring, you plunge your spear point deep into the mammoth just behind the animal's short ribs.

The mammoth stumbles and thrashes; the other hunters close in. More razor-sharp flint spear points flash, and gouts of mammoth blood spurt. You hear the animal breathe its death rattle as you pierce the creature's heart and lungs again.

"Not so bad, old man," you hear your young companion say, as you pant, the air tearing at your lungs.

The mountain of meat trembles where it fell. You lean on your spear shaft, suddenly weak. The years pull you down to the ground. You are called Great Pine and you are aged thirty-nine winters.

≈≈≈

Judging by other unmounted big-game hunting people, the potential life span was probably longer than the actual life expectancy. If a hunter from this period received good nutrition and was not injured, he could expect to live as long as sixty years. The quality of nutrition was not consistent, because it depended on the luck of the hunt and the vagaries of climate and growing seasons. Additionally, hunting gigantic animals was a dangerous occupation; injury must have been a common occurrence. Life was hard, and people aged early.

The material technology of the Clovis people was necessary to their success.

Flint may have been the most important raw material for the paleo people's technology. Flint can be shaped into very sharp implements and can be resharpened quickly. The range of tools that are made from flint include knives and choppers for processing food, drills, wood shaves for making spear shafts and handles, scrapers for defleshing and shaving hides, and of course, spear points.

The creation of the spear point required such a high degree of artisanship that it may indicate their society had specialists who made flint implements. Such specialization implies that sufficient food supply was available so that many people were freed to specialize in a variety of occupations, such as making leather and hide goods, and even religious and/or ritual occupation. If you had the opportunity to watch a Clovis point being made, what might you have seen?

Making the Clovis Point (fictional)

The skilled hand grips the flint nodule in its leather mitt.

"Speak to me, Spirit of Flint. Give me your strong, sharp tooth."

Skillfully snapping the hammer stone down, a piece of beautiful agate flint, about four inches in length, is cleaved from the stone.

"Show me, Oh Tooth, where is your edge."

Antelope antler is applied with force, muscles and sinews strain, and flakes fly away. Large flakes at first, then finer flakes are removed, leaving a straight, clean edge. The point takes shape—longer than wide, triangular and not too thick. The hard labor is rewarded with the excellence of the form. The base of the point is flaked until it is concave and then ground with a stone to make its edge strong. The finishing detail will be the "flutes," two shallow channels, one on each side, running from the base of the point, halfway up the point. The flutes are created with a stiff piece of antler, used like a punch.

"Give me your favor, Oh Tooth, that I may haft you to my spear."

The antler punch is held in precise relation to the base of the unfinished spear point. A moment passes—then the hammer-stone falls. The single flake flies; the flute is formed. The point is turned to the unfluted side and the second blow is struck with confidence. The point is laid aside and the flint core is picked up again, respectfully. You know the hunt will be favored.

The Gault Site

Although excavations at this site are completed, the report has not yet been issued. Still, some of the preliminary results have been released, and they are significant—to say the least. The Gault site has added a wealth of information for interpretation of the Clovis and Folsom Paleo-Indian lifeways.

Paleo-Indians began visiting a unique area in Central Texas, 40 miles north of current-day Austin, at least 12,000 years BP. The site is unique in paleoarchaeology. The Gault site is comparatively huge, perhaps 100 yards wide by 650 yards long. More than 200,000 artifacts have been recovered, and less than a fourth of the site has been excavated. Hundreds of Clovis blades, blade cores, and points have been recovered, as well as animal bones from Columbian mammoths, paleo-horses, and bison. The site was used by Clovis-culture people for thousands of years. All these artifacts might be expected at a site of such age and so admirably suited to use by Clovis people as a base camp.

The traditional view of the Clovis people has been that of a highly mobile society of elephant hunters. These Clovis hunters certainly did range out in their quest for game. The valley location must have been nearly idyllic, but the local game would certainly have been hunted out in short order. There were other attractions in the valley besides game. The valley is very well watered, with an abundant supply of good quality chert. Chert is the necessary material to support the Clovis flint technology.

The traditional conclusions may be overthrown by other findings at Gault. Among the most surprising is the fact that the Clovis blade and point technology continued to be

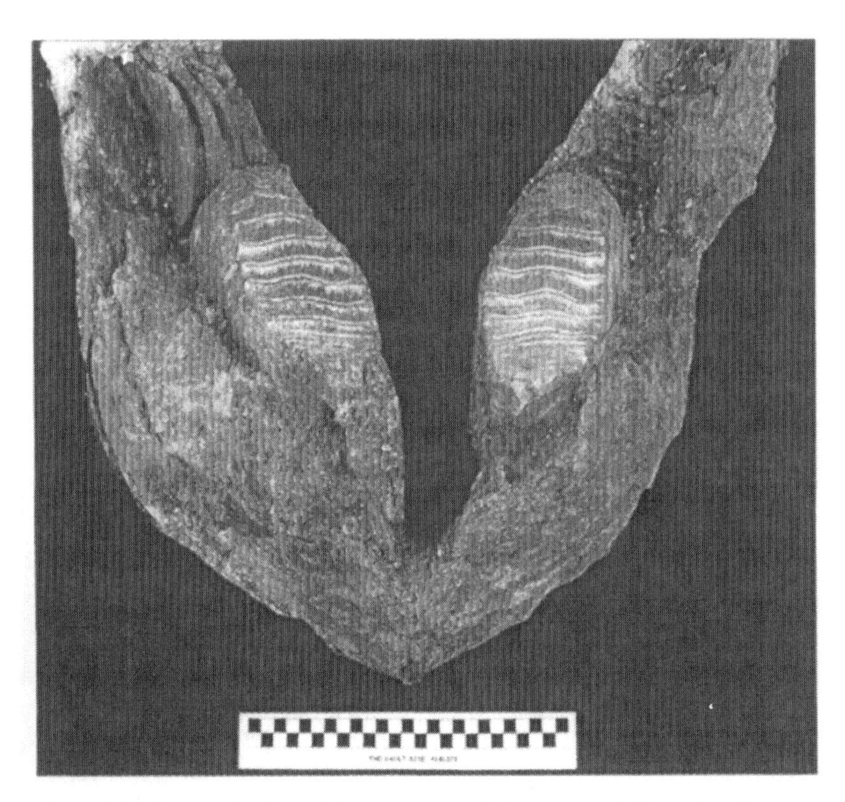

A mandible (lower jaw bone) from a young adult mammoth. Found at the Gault site. Photograph by Milton Bell. Courtesy of the Texas Archeological Research Laboratory, The University of Texas at Austin.

used even after the mammoth disappeared. Many experts had been of the opinion that the Clovis point was a specialization specifically designed to hunt mammoths. The continued use and production of these large, hard-to-make points, well after mammoth extinction, obviously causes reevaluation of that theory.

Small animal bones from turtles, frogs, birds, and small, as-yet unidentified mammals add to the picture. The picture begins to emerge of a generalized hunter-gatherer people, ranging out to hunt big game and making use of small game

Spear points found at the Gault site. Courtesy of the Texas Archeological Research Laboratory, The University of Texas at Austin.

as well. Perhaps the most startling fossil remains found were those of two species of prehistoric horse—thought to be extinct long before human presence in North America.

Other cultural activities have come to light in preliminary examination of artifacts. Clovis blades recovered from the site show signs of micro wear in a distinctive pattern caused by cutting grass. Grass might be useful for bedding or thatching huts. It is certain that Clovis people harvested vegetation for food, and now there is clear evidence for the use of Clovis flint technology for something besides big game hunting. Incised stone with patterns of straight, parallel lines have been found. Some of the stones seem to have figures carved into them, as well. These may be the oldest examples of representational art in North America.

The archaeological investigation of the Gault site began in 1998. The excavations are concluded and the results can now be reported. Current information about the Gault Project may be obtained by visiting the University of Texas Archaeological Research Laboratory website at http://www.utexas.edu/research/tarl/research/Gault/intro/intro.htm.

Summary

The Clovis culture disappeared from North America, and from Texas, after five centuries. Either the various peoples became isolated and individual bands created new refinements to their societies, or a new wave of immigrants found their way onto the continent, bringing new material solutions. The distinction between those who came first, and those who came after, is lost in the sum of those thousands of years since the sojourning began.

Certainly, there is genetic evidence that more than one migration occurred across Beringia. The artifacts that would distinguish between the cultures may have been lost to natural decay. Artifacts of leather and bone, wood and grass, will not survive unless they are protected from persistent forces of water, wind, acidity, and ravenous bacteria. Those artifacts might have demonstrated a clear progression of invention, or a sudden replacement of one people by another. Both patterns happened; the genetic evidence makes that conclusion inescapable. But which was the dominant pattern? We hope for discovery of new sites that will make it clear.

In Texas, the new societies seemed to find just as much success as the previous societies of hunters by becoming more efficient and knowledgeable. By 9,000 years BP, the mammoth, mastodon, the early horse, the rhinoceros, the giant sloth, and all the Ice Age giant mammals were gone from the land. The glaciers retreated, and the Clovis hunters continued to be highly successful with the more modern species. The combination of a changed environment, and pressure from organized and highly skilled hunting ended the age of the megafauna. New skills and technology were required for a new age.

An interview with Dr. Dirk Van Tuerenhout, the curator for anthropology at the Houston Museum of Natural History follows.

The Peopling of a New World: An Interview with Dirk Van Tuerenhout, Ph.D.

Q: Dr. Van Tuerenhout, I had the pleasure of attending a lecture you gave to the Houston Archaeological Society on the topic of the population of the Americas during Upper Paleolithic times. You spoke of several theories of how people might have traveled to the shores of North and South America. I appreciate this opportunity for you to share your thoughts with the readers of *Digging Up Texas*.

First, I'd like to ask you about the possibility that people arrived in the Americas by canoeing or sailing across the Pacific, perhaps hopping from island to island. How likely do you rate this scenario?

A: Given what we know today, I would say that is a very unlikely scenario. If people make a claim that some or all of the earliest inhabitants came from "somewhere" in the Pacific, then they must back that up with data. If those are there, then I would have no problem with that. However, consider the following points, and then see if this scenario still makes sense:

1. We have evidence in the Americas of human presence dating back to at least 13,000 years ago.

2. What do we mean when we say "Pacific"? Australia or Hawaii? That makes a big difference. Australia was inhabited by 35,000 years ago at the latest. Some people argue in favor of making this 50,000 years ago or even earlier. On the other hand, Hawaii was not inhabited until about A.D. 600. If the first inhabitants came from the Pacific, then that would rule out Hawaii but might include Australia.

3. If the first people were to have come from Australia, then how did they cross over? We know that some form of water transportation existed in the Pacific in the Upper Paleolithic. However, look at a world map and consider the enormous distances involved between Australia and any point on the coastline of the Americas. How can we get people there without any archaeological evidence of human presence at that early stage on any of the islands in-between. There are two possible answers here: either we simply have not found the evidence yet, or they never stopped on any of these islands.

I cannot bring myself to buy into the scenario of these early seagoing peoples to have crossed over in one fell swoop. What about provisions? How did they know they had to go east instead of west, or any other direction?

Another point to consider is that any population that arrives on any shore needs to be large enough to be viable, i.e., to have enough members of both sexes to have sufficient numbers of children to make up the next generation. A shipwreck consisting of only men or a very small group of men and women ending up on a shore somewhere do not constitute good founding populations.

A final thought on this topic is why people might have made this crossing. The general consensus about the people crossing over the Bering Strait is that they were following herds of animals. They were hunting and gathering. In other words, they were in search of dinner as they crossed over. The same argument I have heard used to explain how people might have come over from Europe to the eastern shores of North America. What then might people have been chasing across the Pacific in order for them to arrive in the Americas. I must admit that I cannot think of anything.

Q: Does the so-called paleolithic Kennewick Man, discovered alongside the Columbia River in the state of Washington increase the likelihood of arrival via the trans-Pacific route?

A: No. A claim was made that this person might be of Pacific origin. Nothing was offered to substantiate that claim.

Q: Is there evidence for travel across substantial distances of open sea during paleolithic times?

A: Yes. We know that people had to cross a fairly wide body of water to get to Australia. Whenever this may have happened (35,000 or 50,000 years ago), at no time was the sea level low enough to open up a land bridge that would have linked Australia to Southeast Asia. People had to have crossed bodies of water. Moreover, we are very likely not talking about a body of water the width of the English Channel, where, on a good day one can see the white cliffs of Dover from France. These straits were wider, and people could not have seen the shores on the other side. Yet they crossed over. Why? I don't know.

Perhaps they were blown off course when traveling along the coast of what is now New Guinea.

Another piece of evidence in favor of long distance sea travel in Upper Paleolithic times is the presence of obsidian in Japan, at archaeological sites close to Tokyo. This volcanic glass has been retraced to a source that is on an island about 170 miles south of Tokyo. Once more, this island was never linked by land to "mainland" Japan. Yet the obsidian got there and was used by people to make tools with. I can only see one explanation: people had ways of covering that distance.

Q: I'd like to turn your attention now to the possibility of travel to the Americas from the African continent. Some have theorized that carvings found on the Gulf Coast, known as the Olmec Heads, are evidence that Africans made the journey across the Atlantic Ocean and contributed to the population of Central America. How much support do you see for this idea?

A: None. The argument has been put forward that we are looking at African facial features being represented in these Olmec heads. My problem with that is: where is the other evidence that African people arrived on the Gulf Coast? If they made such an impression to have their portraits carved, they must have been there long enough or they must have had enough power to have other things made that would illustrate their presence. As far as I know that has not been found. Still, even if one were to believe that these are portraits of African people, there is this tiny little detail called chronology that would still rule out the use of these heads as evidence that African settlers were among the first to come over to the Americas. While Olmec culture is indeed one of the oldest known Mesoamerican cultures, its earliest dates, around 1200 B.C., do not even come close to the earliest suggested dates for the first Americans. They were around more that 10,000 years after the first people arrived.

Q: Do all the human images associated with Olmec artifacts demonstrate Negroid features?

A: No, they do not.

Q: Another theory put forth in recent years is the idea that people made a trans-Atlantic voyage, or possibly hopped from one ice-bound island to the next, in order to travel from the Iberian Peninsula to the East Coast of the present-day United States. Can you tell us what support there is for this proposition?

A: That is a hypothesis put forward by Dr. Bruce Bradley and Dr. Dennis Stanford. They looked at Upper Paleolithic stone tools in Europe, specifically France and Spain, and found parallels with Clovis tools in the Americas. Chronologically, the European pieces are older than the American pieces, so there is a possibility of logical sequence from an older "mother" culture to a younger "daughter" culture. Crossing over from Europe would have meant crossing the Atlantic. Their argument has been that people might have inadvertently bumped into the Americas while on fishing expeditions along the edge of the ice cap. This is what would have made this feasible: There was food available (fish and bird colonies along the way), and the mechanism is one of gradual reconnaissance of the environment. Perhaps each generation people moved out a little bit more along the ice and eventually came across North America.

Q: The most dominant theory is one of the arrival in North America via the Bering land bridge. Does this one have substantial evidence in its favor?

A: There are various lines of evidence supporting an Asian origin for the first Americans. There are genetic and linguistic data, as well as climatological data that show a land bridge existed. Currently research is being conducted by U.S. universities to compare DNA from Asian and American Indian populations to see what the overlap might be. We do have early sites in Alaska as well. Keep in mind however that this theory does not rule out that people might have come from other areas. This is the most universally accepted and the least controversial among the ideas out there.

This is still a fascinating subject matter. New discoveries are made virtually every year. As new approaches such as genetic research become feasible, perhaps we will be able to answer more questions. As always, once we have found an answer to a question, we will also realize that there are now ten more questions on our list than there were before.

While these discussions and investigations are extremely important we should also not lose sight of what the descendants of the first peoples have to say. They propose that the American Indians have always been here. They know through oral traditions that they have been here since the beginning, and therefore we should not engage in what they consider is irreverent handling of skeletal remains. This approach is one that most non-American Indian people who are interested in these matters are having a hard time coming to terms with. Yet it is part of the picture, and one should include it in the discussion.

Thank you very much, Doctor, for your insights into these sometimes-controversial theories.

Chapter Three

Ancient Material Culture

The material technology (for it was a collection of techniques for the use of different materials) brought to Texas is difficult to discern. What is obvious to even the casual observer is that the techniques changed over time. The materials remained basically the same. Although the stones might be chemically different and the wood, grasses, and canes might be of different species, they were still stones, grasses, canes, and wood.

It is often agreed among archeologists and paleontologists that the flaked stone tool is the most important idea that humankind possesses. Human beings are not well endowed by nature to eat fresh meat. They have no suitable teeth for opening a freshly killed carcass and they have no claws. Humans are well designed to chew vegetable foods and can disarticulate and eat very small animals, but an animal the size of an antelope or small deer has nothing to fear from a weaponless hominid without sharp stone tools.

Scavenging dead carcasses for meat is a viable strategy, and many species use this technique. It has its problems, though. It brings humans into competition with other, better-equipped animals; animals that do have long, sharp teeth and claws. There is the additional danger of disease and illness from eating meat that is too putrid and decayed. Clearly, if humans are going to compete with large carnivorous animals like saber-toothed cats, bears, dire wolves, and lions, they need to do so with technology. Humankind solved this dilemma about one million years BP by flaking African lava to create sharp edges and pointed ends that could pierce tough hides.

By the time the first humans reached North America, they had a sophisticated kit of knapped flint points fixed to long wooden shafts (said to be hafted), as well as stone axes, flint knives and scrapers, awls and needles, and other special tools for processing meat, hides, and bone. They probably made grass and fiber baskets and ropes with great skill. They harvested a wide variety of animal and vegetable foods, and they made clothes and shelters that enabled a few people to expand to a population of millions and dominate two continents in less than two thousand years.

The primary tool of paleolithic people in the Americas began as a long, thin, expertly flaked and exquisitely shaped flint blade. Some purists would say chert blade, rather than flint, but the dispute is academic: Flint and chert are chemically identical stones of microscopic silica crystals. They both come in a variety of colors and exhibit minor differences in texture and hardness. Some experts will declare that the label flint can only be applied to European chert, but the names chert and flint are largely used interchangeably in technical journals and publications. Whatever term is used, the material and the tools are equally fine whether made in the Old World or the New World.

The changes in the paleolithic flint industry in Texas track changes in the rest of the North American continent. The flint blade became something like the "Swiss army knife" of antiquity. In its original form, a long, thin blade, it served one purpose when attached to a long spear and could be used only to stab and cut hide and flesh. This is the classical usage and brings to mind the Clovis big game hunter, stabbing mammoth and mastodon or holding cave bears at bay. Used in such a manner, the blade is highly suitable for the purpose of piercing thick hide and reaching the vital organs of the Pleistocene-age megafauna. Studies of the wear patterns on these blades indicate that they may have been used unhafted, as cutting tools to process food—not necessarily meat—as well.

A spear can be thought of as a weapon system consisting of the payload, the flint point, and a delivery component, the shaft. Many archeologists argue that there may have been a third component as well: a short piece into which the point would be fixed and which would then be inserted into the spear shaft. Whether or not this third element was actually used, it makes wonderful sense. The flint point could be driven deeply into a large animal and the shaft pulled free to be reloaded for a subsequent stab. In addition, the point and short haft could be used as a knife, without the awkwardness of a long shaft getting in the way. The open parklands of central and western Texas would have been a perfect setting for stalking and ambush hunting techniques for which long, stout spears are ideal. A herd of mammoths or mastodons could be driven toward the concealing brush and tree line in which would be concealed a number of hunters. At a signal, several could drive spears into a selected animal.

As the climate changed, so did the vegetation patterns. The climate warmed and became drier. Parklands became prairies, and lakes and ponds dried up. Tall grasses still provided cover for hunters, but the largest of the prey animals disappeared. Mammoth and mastodon were replaced. Herds of deer and antelope took over the browser's niche from mastodon. *Bison antiquus*, the modern *Bison bison*, and other grazers took over the mammoth's niche. A different weapon was needed for a different prey and hunting technique. Spears and their flint points became lighter in order to increase their range. The longitudinal void on each side of the point, called the flute, became longer in order to provide more support for the brittle flint as it was thrown further and with more energy. The additional energy was provided by a throwing stick, one end held in the hand and the other end fitted to the butt of a short spear. This throwing stick, termed an atlatl, (also written atl-atl) increases the effective length of the arm—which is the lever that propels the spear.

The manufacture of flint tools has remained fundamentally the same over tens of thousands of years. Flint (or chert) is a glassy rock that can be found as outcrops, such as those at Alibates Flint Quarry National Park, and as nodules in limestone beds or as river cobbles. Individual types of flint are very distinctive and can often be assigned to a geographic source on the evidence of color, texture, and fracture. Alibates, Edwards, and Arkansas navaculite can often be readily recognized by a novice, but proper identification must take place in the laboratory after careful analysis. Again, the rule of thumb for the avocational archaeologist is if you don't know with certitude, don't give anything a label. This guidance applies generally, with all typology, not just with sources of flint.

When a flint knapper selects a piece for flaking, suitability of shape, integrity (free from obvious flaws and inclusions of other material), and hardness are considered. Three different types of blows are available to the knapper: hard percussion, soft percussion, and indirect percussion. These blows are used to shape the piece into the preform: a blank approximating the size and form of the intended tool. The preform blank is then finished with a fourth technique: pressure flaking. Pressure is applied to the edges of the blank, carefully removing tiny flakes in order to create a thin, regular edge. Careful pressure flaking can create an edge only one molecule thick—much sharper than the sharpest carbon steel edge.

Selecting a flint stone, called a core, for its potential to yield blades or preform flakes, the knapper strikes off an end to provide a platform. If the core has a natural platform—a flat facet that is at approximately 90 degrees angle to the intended flake flat surface—then the knapper omits this stage. Next, a sharp, hard percussive blow is delivered with a hammer stone. The hammer stone is often a river cobble made of quartzite or some other stone slightly less brittle than flint. Flint has a hardness, on the Mohs Scale, of about 7. Pure copper has a hardness of about 3, steel is about 6, and diamond has a hardness of 10.

Sometimes the knapper wishes to produce flint blades or, in their smallest form, microblades. The flint blades are made with one hard percussive blow. Flint fractures concoidally: the energy of the blow traveling in concentric waves in the form of a cone. The cone can be controlled by the direction and force of the blow, actually becoming very flat and directional. This property is valuable for creating thin, flat pieces. The flake that makes a blade is long and thin with very sharp edges, longitudinally. It usually has a longitudinal curve similar to an animal claw, or canine. It usually has a central ridge down the length, called a crest. Blades are very common and are thought to be the prehistoric version of utility knives. They are quick to manufacture and make efficient usage of flint through providing the maximum amount of cutting edge with the least amount of core consumed.

Points, knives, and scrapers are much more difficult and wasteful of core material but they provide additional functions and capabilities not available with blades. These items can be either unifaced, knapped on one side only, or bifaced, knapped on both sides toward the center line. The flake from the core is reduced through direct percussion with blows from a hammer stone or billet. A billet is a hammer made of bone, antler, or wood that provides soft percussion. Thinning and shaping flakes are removed and readies the work for the final stage: pressure flaking.

In the Folsom point, we see the highest art of the flint knapper. The final stage requires the removal of two longitudinal flakes, one on each side, to form the long, shallow, even flutes. After the shaping and sharpening of the point with pressure applied by use of antler tines, and the final retouching to create the exact shape and aesthetic quality, the point is ready to acquire the flutes that make it a Folsom. I've spoken with modern-day knappers about this crowning feature, and they consistently tell me the same thing: After some time, ranging from 30 minutes to two hours, they achieve the final form and are able to remove the first flute. But the point breaks with the attempt to make the second flute. This is the case whether the flutes are made with direct or indirect percussion (striking a billet with a hammer) and even when using strong direct pressure. The creation of the second flute remains a closely guarded secret by those who've mastered it, or it is truly a matter of fortune.

The Folsom and Plainview People: Paleo Prairies

Over the course of the next few centuries after the Clovis period, the last glacial ice sheet of the Wisconsin Ice Age melted and receded to the north. The winters were still cold, but the summers were warmer and lasted longer. Wet spring and summer seasons gave way to a dryer growing season. On the open parklands of West Texas, trees became scarce and tall prairie grasses replaced them. New fauna gradually replaced the giants of the Ice Age. With the experience of more than four thousand years behind them, the people of Texas were sophisticated in their knowledge and technology and were able to exploit the resources of their environment. Burgeoning population and changing climate required new efficiencies and better solutions for the problems of everyday life.

Some of the megafauna persisted for a few thousand years. *Bison antiquus*, about 15 percent larger than the modern American bison, replaced mammoth, mastodon, and camels as the prey of choice for the prehistoric people of Texas.

Bonfire Shelter — Langtry, Texas

The oldest evidence for mass kills in the Western Hemisphere is found below a cliff near the West Texas town of Langtry. The tall cliff abruptly interrupts the vast desert, which was a tall grass prairie at the time of the Folsom people. The open parklands of the glacial periods receded as rain became less frequent. The climate change had its effect on the animals inhabiting the area. Modern *Bison bison* were seen in great numbers as the ancient *Bison antiquus* disappeared. Deer and antelope replaced the mastodon and mammoth that browsed in the open woodlands and nearby grasslands shortly before.

The Bonfire Shelter site shows layers of bison bones left from ancient hunts. The herds were driven to the edge of the cliff—then over it. The resulting tangle of dead bison were butchered on the spot. The earliest layer of bones belong to *Bison antiquus*. The remaining layers of bones are all of modern bison.

The processing of the animals was apparently done in the villages; only butchering sufficient to reduce the size of the pieces was carried out at the kill site. The process was

One possible technique used by Folsom culture bison hunters is depicted here. The animals depicted are *Bison antiquus*, but the same technique would have worked with the modern species, *Bison bison*. Drawing by Jason Eckhardt.

routine. Evidence suggests that the front legs and haunches were removed first, then the side and belly meat was taken. The fatty hump of the bison and the marrow bones were prized by Native Americans in later times, so it is logical to assume they were highly valued by these ancient people as well.

Villages and Camps

Back at the village, the women would have processed the meat further. The fact that large amounts of meat were obtained in the hunt indicates they preserved it. It may have been smoked or dried. It may have been processed further by chopping or grinding it and mixing the protein-rich pulp with fat or meal made from ground mesquite beans.

Fresh meat and bone marrow might have provided a feast for the first few days after the hunt. The organs of the bison would have been collected. Later people often ate liver, kidneys, and hearts raw, savoring the variety in their diet, and so might have these people.

If the food was cooked—and certainly much of it was—it would have been roasted or boiled. Roasting would have been done over an open-hearth fireplace. Boiling required a bit of ingenuity, since these people did not have pottery. Large stones found near hearths give evidence of a practical solution: Tightly woven grass or other fiber baskets were filled with water. The rocks were heated in the coals of the fireplace then dropped, red-hot, into the water, along with the food. Hot rocks could be added to keep the water boiling as long as needed.

Cooking the food allowed the human digestive process to obtain much more nutrition than did consumption of raw food. Larger populations could have been sustained by smoking, aging, and cooking meat. As populations became larger and animals became smaller, this must have been a vital technique for the people of this time.

Gathering and foraging might have become much more important during the Folsom-Plainview period. Food items (mesquite beans, pecans, cactus pads, and fruits, wild onions, roots and tubers—each in their season), materials for weaving baskets and clothing, and medicinal plants, to name a few items, would be required for a people to whom processing and storage had become necessary.

The villages were semipermanent and utilized on a seasonal basis. The villages were placed alongside streams, near bison migratory trails and other resources. They may have been as large as fifty families or more. It is likely that they were very isolated from each other, as the evidence for them is scarce. A village would be abandoned at the appropriate season, and grinding stones, hearths, and any other item too heavy to carry would simply be left in place for use upon the return of the villagers.

At some point during this period, dogs began to be utilized as pack animals. When they were domesticated is not clear in the archaeological record, but it is well within reason to suppose they played a part in human society by this period. Temporary campsites from this period suggest that the village group did not make its way en mass, but may have traveled as individual family groups, each family foraging and hunting along the way.

Artifacts

Few Folsom and Plainview artifacts have survived from this period of human endeavor. Campsites can often be dated to this period through the use of carbon-14 radiologic dating, giving some idea of the range and distribution of the culture. Through the information such artifacts and traces provide, it seems evident that the Folsom-Plainview peoples were concentrated in Texas and did not range widely beyond the area.

Folsom Point

The spear points found at sites from this period reflect a change in prey. The Folsom point found at bison kills is smaller, with a flute that runs nearly the full length of the point. The long flute on both faces may reflect a different method for hafting the point to the spear shaft. Some paleontologists, specialists in prehistoric life forms, speculate that the Folsom point may have been fixed to a wood or bone foreshaft, which then could be attached to an atlatl shaft. This technical modification would allow a spear with a broken point to be quickly equipped with a new spear point and put back into service without interrupting the hunter's participation in the kill.

Scrapers, Knives, and Other Stone Tools

While they often stimulate the imagination most, projectile points provide the least part of the record of human activities in the dim past. Scrapers and knives, being essential to preparation of meat and hides, are far more abundant at campsites. These necessary tools were made much more easily than were the carefully crafted points, and were not kept track of as carefully. A scraper or knife might be made on the site of a temporary hunting camp and discarded after use. Archaeologists have a term for careful preservation of goods: *curation*. An item that is rare, or one that takes a great deal of expertise to make, will be a highly curated item. On the other hand, a fist-sized stone that is good for heating in the coals of a fire and used to boil a skin-full of water would not be valuable. Fist-sized stones are too abundant to be worth the effort of packing them around. There is little doubt that many highly curated items are under-represented in the archaeological record. Still, the common household tools can tell us a great deal about everyday human activities.

The tool kits change little over time, in terms of the kinds of tools, but they change significantly in other aspects. As large animals decrease in number and smaller animals are hunted instead, the tools used to process them become more important to the bands of people relying on them. Smaller animals with thin hides don't require massive, hard-to-make Folsom points. They do require sharper, fine-edged scrapers to remove the thinner layer of fat from the hide. Good knives might become more important than finely made projectile points because a small animal is easier to kill, but it takes much more processing time to skin and butcher the number of rabbits that equal the amount of meat on one mastodon.

In addition to smaller game, nuts, fruits, tubers, and edible plants become a more important dietary component. Tools like *denticular blades* become more abundant. Denticular blades are small "tooth-like" flint and chert cutting tools. They can be used individually, or they might have been set into wood to make a rudimentary scythe. Choppers and grinding stones are relied upon to crush and pulverize acorns, mesquite beans, and grains.

Lake Theo Folsom Bison Kill Site

Stone tool fragments were discovered in 1965 and 1972 on the shores of Lake Theo, named for former landowner Theodore Geisler. Archaeological testing in 1974 revealed a campsite and bison butchering and processing area dating back to the age of the Folsom Man, between 10,000 and 12,000 years ago. Projectile points and scraping tools were found at a depth of four feet, along with over 500 bones from an extinct type of bison. Ponds in Holmes Canyon now covered by Lake Theo probably attracted prehistoric man to this area.

Lake Theo Folsom Bison Kill site, Caprock Canyons State Park and Trailway. Bones of the extinct *Bison antiquus* were excavated at this site. Artifacts found in association with the bone bed indicate extensive use of this area for butchering and processing bison meat over several thousand years. Photograph by Robert Marcom, 2001.

The dry lake bed at Lake Theo Folsom Bison Kill site is bounded by 40- to 50-foot cliffs. Photograph by Robert Marcom, 2001.

State historical marker: Lake Theo Folsom Bison Kill site. Photograph by Robert Marcom, 2001.

From the historical marker: "A road older than recorded history, carved out in centuries of wintertime travel to the south, spring migration to the north by millions of bison and by Indians who lived by hunting these large animals. Important in era of Texas Panhandle settlement. Used in 1873-1874 when first lifelong residents put dugout dwellings in the Panhandle and began to hunt buffalo to fill demand for hides and meat." Photograph by Robert Marcom, 2002.

Old Mobeetie Trail historical marker. Photograph by Robert Marcom, 2002.

About 9,000 years ago, Folsom age people drove hundreds of *Bison antiquus* off the far cliff to their deaths. Just as the people of the Bonfire Shelter did, these ancient prehistoric men and women skinned and butchered the carcasses where they lay. Numerous tools, including specialized scrapers, points, grinding stones, knives, gravers and abraders were found alongside and among a bone bed between 6 and 14 feet thick buried under 4 to 6 feet of red sand overburden. Two distinct layers of Pleistocene era bison bones were carbon dated. According to *The Handbook of Texas Online*, "a single carbon-14 date of 9,360 ± 170 BP was obtained on bone samples from the Folsom bone bed. A second test on bone associated with the Plano points provided a date of 8,010 ± 100 BP."

Dr. Jeff Indeck, curator of archaeology at the Panhandle-Plains Museum, Canyon, Texas, sums the importance of the site up succinctly:

"Lake Theo represents a multicomponent archaeological site. There is evidence for Paleoindian bison butchering 9,000-10,000 years ago. A Folsom projectile point was found along with a possible ceremonial altar, in association with an extinct form of bison. Foragers (hunters and gatherers) with large dart points and scrapers occupied the area about 2,000-3,000 years ago, perhaps during journeys to exploit seasonal food. A possible ceremonial altar was recovered at Lake Theo. It consists of bison bones arranged vertically within the bone bed. It is possible that this feature was topped with a bison skull, similar to cairns erected by medicine men over historic bison sites and documented by observers of Northern Plains Indians."

The bison that died at Lake Theo followed an ancient migratory path, called the Mobeetie Trail, along a system of valleys following current day State Highway 70. Migrating south in the winter, the bison found grass and water on the southern plains, and the Folsom people knew where the herds would travel. They may have hidden in the tall grass as the herd flowed around them. The hooves of the bison in the herd would have produced a constant low rumble and trailed a cloud of thick red dust behind them as they walked. Rising up at the strategic moment and waving hides, shouting, and charging, the hunters would have frightened a portion of the herd, stampeding them toward the cliffs.

The hides were removed, and the fat and flesh were scraped from them. Selected portions of the carcasses were cut from the bison carcasses and possibly carved up and distributed on site. There is also evidence of ceremony, showing a religious or spiritual aspect of the Folsom culture. Fossil jawbones, lower leg bones, and the tall spine of a backbone were found at the bottom of the deposit, standing upright in a circular hole.

The site was not occupied again for thousands of years. Beginning about 3,000 years BP, the site was again utilized for butchering, this time of the modern species *Bison bison*. The points, tools, and implements were typed and dated to the Archaic period, apparently deposited between 1000 B.C. and A.D. 500.

Chapter Four

Archaic to Proto-historic Cultures of Texas

Archaic Material Culture

The materials and industries that became important to the Archaic people were not all invented by them. There is little doubt that previous cultures wove fiber into mats and baskets. Returning to a locality in time to reap a harvest of nuts, fruits, or beans is an idea that preceded Archaic cultures, as well. Baskets and harvesting are both ideas that had value to previous people and their lifeways, but they became especially important at this period of human development. Populations were growing and the need for a predictable source of food in every season was a matter of the survival of the tribe and the bands of which it was composed. If the bands were not strong and healthy, then the tribe would not be able to protect its territory. If the tribe lost its territory, then it would cease to exist.

Harvesting was probably a much more organized affair than gathering. If a band was interested in the harvest provided by a particular grove of pecans, then it is likely they would plan to camp in the vicinity and collect as many pecans as possible. They would use baskets to contain the harvest and stay in the area until the harvest was exhausted. Archeological evidence supports this idea. Small open campsites from the Archaic era give evidence of frequent abandonment and reoccupation.

Implements for food processing are frequently found at Archaic campsites. The need to grind mesquite beans and later, maize, resulted in the *mano* and *metate*. Manos (grinding stones) and metates (stone basins) are quite heavy and were probably left behind, rather than carried from campsite to campsite. Another method for processing included the use of pits, both naturally occurring and man-made, along with heavy sticks for pounding material placed into the pit.

The invention of bow and arrow came at the end of the Archaic era. Hunting (and gathering) continued to be a major occupation for Archaic people. The atlatl is an unwieldy weapon for use on small game and probably does not have the effective range needed for hunting a deer and antelope population grown wary through long experience of being hunted. Dates for the earliest use of bow and arrow in Texas are largely speculative. They range from as early as 2,500 years BP, to as late as 1,500 years BP. The bow and arrow has been invented by hunting cultures all over the world. It is a relatively late invention in the Americas, perhaps because of the abundance of game and because prey in the Western Hemisphere evolved without the presence of human hunters. In such a circumstance, the atlatl might be sufficient for a very long period.

Texas Archaic Cultures

Texas is an archaeological phenomenon due to its unique attraction to humans over the past 12,000+ years. During the late Pleistocene, it was open parkland and tall-grass veldt with now-extinct megafauna like mammoth, mastodon, *bison antiquus*, giant sloth, and many more. Newly arrived humans with mongoloid features (phenotype) found the hunting good and the resources spectacular. The founding cultures may actually predate the oldest one universally accepted: Clovis Man.

Native Americans stayed on in Texas after the end of the Ice Age and adapted to the dramatic shifts in climate and prevalent species. Some, like the Apache, became desert specialists. Others, like the Cheyenne and Arapaho, continued

Activities depicted include a successful hunt, flint knapping, grinding mesquite beans, and preparation of cactus pads for food. To the right, meat strips are being dried.

The dwellings are comprised of poles lashed together and thatched with grass mats. The weapons carried by the hunters are atlatls: a short spear used with a throwing stick that acts as extra leverage by lengthening the effective distance from the hunter's shoulder to the spear haft. The spear can be thrown with nearly double the force as compared to one thrown without the atlatl. Drawing by Jason Eckhardt.

to make a living on the plains, hunting big game. The early Mississippian culture learned to farm and harvest the eastern forests, increasing the yield to feed burgeoning populations.

The Native Americans' world was in constant turmoil, as the Indian nations outside Texas's boundaries attempted to expand into the already-occupied areas. The pressures from population growth, changes in resources and climate, and extraordinary competition for the same limited resources provided high motivation for conflict. Warfare, offensive and defensive, became a necessary art.

This must have been especially true in the last years of the previous millennium. About 1,000 years before present, the struggle becomes apparent in the archaeological record. It lasts to the time of contact with the French and Spanish (and some believe, with Anglo-Dutch pirates) in the middle sixteenth century.

Six thousand years ago, Texans participated in a worldwide cultural change. Even though they were isolated from groups on the other side of the world, they entered the Archaic period at about the same time as their distant kin. Why this happened puzzles archaeologists. One major clue to why this change occurred is found in the record of climatic change.

The Wisconsin period of the Ice Age ended about 9,000 years ago. Seasons changed; longer, warmer, and dryer summers became the rule. Less rainfall meant fewer, smaller lakes. Full-time running streams were further apart, and small isolated *biomes* (sustainable biological systems of animals and vegetation) developed along them. The Ice Age megafauna were replaced by smaller animals. The parklands of West Texas disappeared and were gradually replaced by short-grass prairies.

Whether the much-romanticized big-game hunters of the Ice Age were put out of business by their own skillful hunting to extinction of their megafauna prey, or whether the climate killed the big game animals off is debatable. The fact is, they disappeared, and new food sources replaced them. It is important to say that these dietary changes are not differences of kind, but rather differences of quantity.

Human beings are and always have been omnivores. Bands of people always varied their diet with ripe berries and nuts when the opportunity presented. It is unlikely that people would refuse a convenient meal of rabbit stew for the adventure of the mammoth hunt. Vegetables have always been nutritious, and paleolithic people surely would have noticed that. Modern anthropological studies, called *ethnographies*, indicate that historically recent hunter-gatherer societies also collect and store material resources as they need to do so. They will forage for food in times of abundance, moving their camps as necessary. When drought or disease reduces game, they will settle near relatively good seasonal food sources and wait out the scarcity. The idea that any culture at any time was entirely characterized by big game hunting is probably mistaken.

In the Archaic period of Texas Native American habitation, the balance of activity changed in favor of occasional bison, deer, and antelope hunting, settlement near rivers, ponds, and full-time streams, and more time spent gathering and collecting. Projectile points became smaller and less elegant as a trend—though many of the smaller points are works of art—and other tools became more important to bands of people than before.

How it Might Have Been

Gray Mouse loved the summer. He knew the summer was a good time for the boys of the tribe. Other than the occasional deer the warriors brought home, meat was provided by the older boys. Not yet burdened with the responsibility of adulthood, the boys were free to combine play, sport, and small-game hunting and to sample the rewards provided by an appreciative tribe.

Only one thing worried Gray Mouse this summer.

"Are you sure? Why didn't you tell the Council you saw them?" Gray Mouse studied his friend Hopping Crow carefully. Hopping Crow was a good friend but not always a smart one.

"I always tell you first. Should I tell others?"

"Let me hear it again. What did you see?"

"By the big marsh, on the far side, I was killing frogs. You know, you must be very still in the dusk and wait until you hear them croak. I was sitting in the water with my frogging stick, when I saw another boy. He was not from this tribe. He had a frogging stick, but different from ours. He wore no leggings, but only a loincloth. He was not a good hunter because he walked along, stabbing at the water."

"Why didn't you follow him when he left?"

"Follow him? I didn't think of it. I only thought to come to tell you. Should I have followed him?"

"You did fine. We will see if he comes back tonight."

The boys kept their secret to themselves that evening. Gray Mouse walked past the drying racks, where deer and antelope meat was being cured, down the short trail to the place where the Flint Maker worked. Stone Eye was sitting on his short stool, working carefully on a ceremonial knife for the tribal chief.

"I see you, Gray Mouse. Sit here and tell me what is new."

"I see you, Stone Eye. You still have not finished your knife. How does it go?"

"Perhaps I should have stopped sooner. I have a problem, but it is one of my making."

Stone Eye paused, noticing his question went unanswered. "Do you need something, little man?"

"I would like a sharp flake, like you make for Black Bird Woman."

"I see. You wish to slice meat?"

"Perhaps to slice meat. But I will tie it to a stick, so make one end dull."

"This meat you would slice—you need to reach it with a stick?"

"Perhaps I need to reach it. I have a new idea. I think I would like to try it before I tell anyone. One knows new things can bite like the coyote."

Stone Eye knew the term. Bite like the coyote meant the thing might be tricky and make Gray Mouse look ridiculous. He respected such caution.

"I have such a flake. Look through that pouch—not the deer pouch, look in the old buffalo skin—it is curved and thin. No, not the large red one. I knew you would like that one, but it is not for you. The thin brown one—do you see it?" Stone Eye divided his attention, watching Gray Mouse carefully. A gift could not be taken back, even if given mistakenly.

"Good. Bring it here." Stone Eye took the flake and ground its base with a small, worn piece of dolomite.

"Thank you, Stone Eye. I will tell you of my idea if it works."

"Be careful of the meat you will cut. Some meat will cut you back."

The sun arced across the sky as it did every day. To Gray Mouse and his four companions, it seemed to take much longer today. Gray Mouse took his flake and fashioned a spear. He found an old lodge pole, one that had cracked, and broke it. Using the flake carefully, he split the broken end of the pole then inserted the ground end of the flake. He intended to capture the young foreigner, and he needed a weapon to threaten him. He did not wish to kill the foreigner, and so he needed the help of the other boys.

They fell in with his plan, because adventure was not common in this small village, and they thought they would be rewarded with fame and accolades if they could carry out this daring deed. Dusk finally came, and it found the four boys concealed in the reeds. Just as Hopping Crow had said, the boy came, walking upright in the shallow water, and poking at the marsh with his stick. The boys had chosen their spots strategically, and the foreign boy found himself surrounded and under threat from Gray Mouse's spear.

The boys marched their prisoner through the marsh, up the trail, and into the camp. They imagined they saw admiration on the faces of men, women, and children well known to them as they passed the lodges. They passed into the center of the small village, pushing their captive ahead of them, and made their way to the council fire.

Gray Mouse stepped in front of the procession. He drew himself up and spoke.

"I bring an enemy of the people before you, oh Council!" The moment of triumph shined in his eyes, and his companions reflected his glory with their confident smiles.

The gray-headed man seated at the fire looked up—puzzled at first, then with a faint smile growing on his face.

"I see we have budding warriors in our midst. Tell me, brave ones, where did you take this enemy?"

"We found him stealing our frogs from the marsh!" Gray Mouse cast a ferocious glare at the foreigner along with the accusation.

"And, where are these frogs that he stole?"

"We gave him no time. He was prevented from his theft because we stopped him."

"I see. You, young 'enemy,' do you speak for yourself?"

"I stole nothing. I hunt no frogs. I take only fish, as we have been told by our chief. I take only fish, for that is what we have agreed with you."

"Well spoken, young 'enemy.'" The chief turned his gaze to Gray Mouse and his companions. They shifted about, uncomfortable with this turn in the conversation.

"You little men should have come to the council first. This is a boy from the tribe of Fish People. We have agreed that they will take fish from our marsh, and they will give us shafts of wood from their spear trees. This is no enemy. If you had asked, you would have been told. Still, I see we will have fine warriors and brave hunters from you. Now, show your bravery. Take this fish hunter through the Night Spirits and back to his people."

≈≈≈

Gray Mouse lived in a world very different from that of the Paleolithic people. Larger populations became possible through more intensive foraging and by a more efficient use of resources, while suitable areas for settlement became fewer and more condensed. The eastern woodlands were less obviously affected than the plains, but the same trend brought about changes there, as well. Scarcity is often the spark of human innovation, and both the people of the woodlands and of the plains were innovative in their response to their environment. Later in the Archaic period, baskets would be replaced by pottery in societies where villages were permanent. Agriculture would replace foraging. And collecting and storing food would become necessary where seasonal migration became impractical. As populations became larger, territorial resources would be defended. Even nomadic tribes, which would persist into the Historical period, would be constrained by militant opposition from other tribes.

Querechos and Teyas: The Plains People of the Archaic Period

The last few hundred years of the Archaic period, from about A.D. 1300 to 1540, are known as the *Proto-historic* period. The Native American tribes and nations of this time period would become known by their historic names to the Europeans and Americans who would "discover" them: Apache, Pueblo, Tonkowa, Wichita, Comanche, Karankawa, and Caddo to name but a few.

The Spanish expedition led by Coronado in A.D. 1541 identified three groups of Native American people in the western plains and Panhandle areas of Texas as the nations of the Querecho, the Teyas, and the Jumanos.

The Querecho and the Teyas are identified in the archaeological record as distinct cultural groups who inhabited the West Texas plains and the Panhandle area after A.D. 1300. The Querecho would later be known as the Apache. It is likely that the Teyas were an offshoot of the Plains-Caddoan culture—possibly identified by the Spanish as Jumano. The Teyas are known as a separate people only because of the record left to us by the Spanish explorers. The Teyas and the Querecho are difficult to distinguish by artifactual evidence alone. These historical records allow archaeologists to construct distinct ethnological profiles for these two peoples.

Archaeological evidence demonstrates these two peoples had much in common: Both were bison hunting cultures, both lived in seasonal villages called *rancherias* by the Spanish, and both participated in commerce and trade with the Pueblo people far to the east. Evidence for Puebloan trade indicates they traded bison meat and pelts for pottery, turquoise, obsidian, and marine shell beads.

The Tierra Blanca and Garza Complexes are the primary sites for the archaeological evidence of these peoples. The complexes lay on both sides along the headwaters of the Red River, in the western Panhandle. The complexes include the western part of the Llano Estacado (Staked Plains) and the Caprock Canyon formations to the east. Both complexes include residential and base campsites, bison kill sites, rock shelters, and burial sites. They used tipis for shelter and were nomadic. They lived in small, scattered camps during the summer and moved to large encampments during the winter months.

A central question regarding Plains-Puebloan trade is why trade at all? The Pueblo people were settled farmers who grew their food, and who had access to trade with wealthy Central American cultures. Why did they want bison robes and meat? The Plains people were nomadic hunters for whom surplus goods were inconvenient. Why did they want pottery? The Plains people had access to mesquite beans, roots, fruits, and berries, so why did they want maize?

The Antelope Creek People—Puebloan Culture in the Prehistoric Period

The Antelope Creek people occupied the Texas-Oklahoma Panhandle area after A.D. 1200. They were semi-sedentary farmers much of the year, but prepared to follow the buffalo when low rainfall made it necessary. These people built stone and wood houses, made puebloan style pottery, and invented a wide variety of specialized tools.

The Antelope Creek Project opened in February 1938 and closed in January 1939. The principal investigators (PI) were Ele M. Baker and Jewel A. Baker. Of the eleven months duration, six months were devoted to excavation at the site then known as Antelope Creek, Ruin No. 22, and the remaining five months at Alibates, Ruin No. 28. The report filed by the two PIs state that 40,000 cubic feet of dirt was excavated in this eleven-month period. (*Archaeological Excavations of Antelope Creek Ruins and Alibates Ruins Panhandle Aspect 1938-1941*, by the Panhandle Archeological Society, Publication No. 8, 2000.)

The Alibates Flint Quarry, now a national park overlooking the Canadian River and adjoining the Lake Meredith Recreational Area, is comprised of rolling prairie with outcrops of a high-grade agatized dolomite. Geologically, the dolomite rock, a hard limestone with magnesium, has become permeated with silica in solution and the end result is large outcrops of chert (same as flint) with a very fine crystalline structure, ideal for flaking into lithic tools. The chert has distinct physical and chemical characteristics and has been found to be the source material for lithic tools found over a large range of Central and Western North America. The Native American people who lived in the Alibates area from A.D. 1200 to 1350 were the Antelope Creek Puebloan Indians.

These people farmed and traded, much as did the other Pueblo Indians. They hunted buffalo. They built substantial houses from limestone and/or dolomite and used wood poles for upright frames and roof joists. They had an elaborate ceramics industry, which allowed for relative dating of the occupation. Pottery types used for this purpose included "Augua Fria Glaze on Red, Cieneguilla Black on Yellow, and St. John's Polychrome."

More than fifty-three rooms were excavated between the two sites (including three rooms at Antelope Creek, Ruin 24) and more than 750 identifiable artifacts were recovered. The rooms were rectangular with post holes evident. They had fire pits in the floors, channels cut into the clay floors along the walls, and low entrances facing east. The walls

These houses, built by the Antelope Creek Puebloan people, are ideally suited to the climate of the High Plains. Drawing by Jason Eckhardt.

typically were constructed of double slabs of tabular (flat) stone stood on edge, filled between with earth and cemented with red clay mortar. Intriguingly, one small circular room was identified at Alibates, Ruin No. 28. Its function is unknown. It had no fire pit and contained no artifacts.

Artifacts found included bison bone hoes and gouges, bone awls and needles, knapped chert projectile points, knives, drills, axes, choppers and scrapers, pottery (restorable ollas and fragments), quartzite manos and metates (grinding stones), and ornaments of bone, shell, and turquoise. The projectile points were often of the side-notched

This stone hoe blade would have been used by hafting it to a wooden handle. Also shown is a stone abrader, possibly used to straighten arrow shafts. Photograph by Robert Marcom, 2002. Courtesy of White Deer Land Museum, Pampa, Texas.

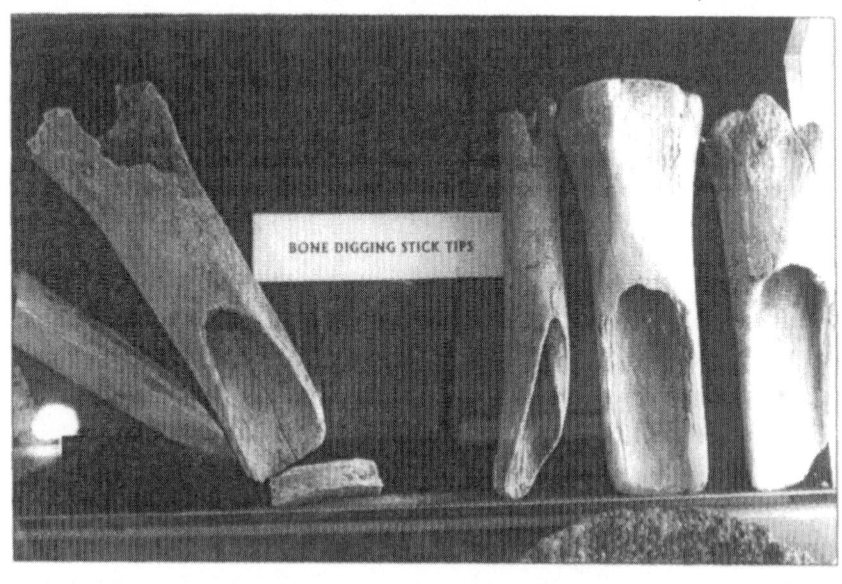

Bone artifacts, which were probably used to dig. The label, *stick tips*, suggests that they would have been used with wooden handles. Photograph by Robert Marcom, 2002. Courtesy of White Deer Land Museum, Pampa, Texas.

variety, but several were triangular with convex or concave bases. A few corner-notched points were found and were judged to be intrusive, deposited by nomadic Indians.

Fourteen burials were found, typically near the northwest corner of a room, inside or just outside the walls. The bodies were buried with knees flexed and often lying on their right side. Several of the burials had slabs of dolomite placed over the bodies. Burials were comprised of both males and females and ages ranged from infants of a few months through adults of 40-50 years of age. No grave goods (material items buried with the bodies) were noted in the report for the Antelope Creek Ruins. Two of the burials at Alibates, Ruin No. 28 included mortuary (grave) goods: shell beads (child c. 10 years); a bone awl (an infant).

An interview follows, with Jeff Indeck, curator of archaeology at the Panhandle-Plains Historical Museum located in Canyon, Texas. The Panhandle-Plains Museum offers an extensive collection of historical items, artifacts, and their interpretation, including those of the Antelope Creek archaeological excavations.

Interview with Jeff Indeck

Q: The museum has an extensive exhibit and interpretation of the Antelope Creek people. Unfortunately, some of the site was destroyed by construction before it could be documented. Do we have a good picture of who these archaic Native Americans were, and about their cultural activities?

A: Archaeological evidence for Antelope Creek culture is moderately well documented. Although vandals have indeed damaged the site at Antelope Creek, numerous other sites in the region, occupied during the same time, provide fairly detailed information about Antelope Creek activities. These farmers lived along tributaries to the

Canadian River approximately 550 to 850 years ago. They lived in houses made with stone, sticks, and mud; they hunted bison; raised crops of corn, beans, and squash; mined Alibates flint; and traded extensively with people to the west and northeast.

Technically, Antelope Creek people do not reflect an "Archaic" strategy for making a living. An Archaic lifeway reflects foraging (hunting and gathering) and is characterized by exploitation of many foodstuffs and use of large corner-notched projectile points. Antelope Creek represents a regional variant of late prehistoric, Southern Plains Village culture and is typified by cord-marked pottery, small side- and corner-notched points, bison bone digging sticks and hoes (horticulture), and artifacts made from Alibates, which was quarried from pits in an area south of the Canadian River near Fritch, Texas.

Q: Is there evidence for trade networks or for contact with other Native American cultures?

A: The best examples of Antelope Creek trade and contact with other people include numerous exotic (non-local) materials like seashell, turquoise, and obsidian, which were probably acquired through trade. Antelope Creek houses reflect many characteristics similar to pueblos and may indicate contact with people to the west and southwest. Evidence for Pueblo-Southern plains Indian contact is represented by pottery for approximately two thousand years. The roof structure and some of the house layout, along with pipestone and Niobrara jasper, may reflect contact with groups to the north and east. Contemporaneous groups along Wolf Creek and in the Oklahoma Panhandle certainly had extensive ties to other plains groups, probably including Apaches. Also, most researchers believe that Antelope Creek people were incorporated into either the Wichita or Pawnee, which further indicates close relationships with northern plains people.

Q: How did these people differ from the nomadic people of the plains?

A: The most significant difference between Apache and Antelope Creek Indians is that Antelope Creek people lived in semipermanent houses of mud, stone, and sticks, and practiced horticulture (farming without draft animals). Early Apaches were bison-hunting foot nomads living in hide structures and employing dogs as pack animals. We believe these are the people chronicled as Querechos in the narratives of Coronado and Oñate.

Q: Are these Native Americans comparable to modern people in their needs and in their uses of the environment?

A: The needs of people really have not changed much during the past 14,000 years. Water, food, shelter, and clothing have always been necessary for survival in the semiarid environment of the southern plains. When resources are not available in the environment, people can either move to the resources or trade for them, and we see examples of both behaviors across time. Interestingly, as we show in our new People of the Plains exhibit, there are very few activities today that do not have analogues in the past. For instance, water has always been in short supply. When Euro-Americans arrived during the late 1870s, they settled areas around springs and playas, the only surface water available. With the introduction of the windmill, people dug wells to obtain water from underground aquifers. Many archeological sites are also concentrated around springs and playas, and there is extensive evidence that prehistoric Indians dug wells to search for underground water as early as 13,500 years ago. The earliest inhabitants of the region "mined" grass by hunting mammoth and bison. Today ranchers herd cattle, but the end result is meat from grazers eating the grass on the southern plains. The region has no brick and few trees for making houses. Consequently, early houses were dugouts, pickets, and soddies made from available material. There is very little difference between a historic dugout and the houses of Antelope Creek people. Also, even today, a few underground houses are used in the Texas Panhandle, and these houses are energy efficient and able to withstand the damaging effects of hail and tornadoes.

Chapter Five

Woodlands Material Culture

The two greatest contributions of the Late Archaic Woodlands phase prehistoric Indians are agriculture and ceramics. The Texas Caddoans raised both occupations to the level of a specialty. Horticulture required special tools, such as hoes, to work the ground, as well as tools and utensils to process the harvested food. The hoes were made of stone and fixed to wooden handles. Manos and metates were inherited from the Archaic era and were refined by the Woodlands culture. Manos (grinding stones) and metates (stone basins) were used to grind maize kernels, nuts, and dried meat. Special choppers were knapped from flint, appropriate to specific foodstuffs.

Two forms of agriculture were practiced: horticulture, which is dry-land farming and dependent on rainfall, and gardening, which is watered by transporting water to the crops. The three staple crops of Native American agriculturists are maize (Indian corn), beans, and squash or pumpkins. These staple crops could be grown in either fashion, but horticulture is completely dependent on the rains coming on time and in season. If the water source was too distant for transporting water, then horticulture would be the only choice available.

The tools that were designed and produced specifically for agricultural use enabled more efficient cultivation, offering an increase in the size of the crop and therefore supporting larger populations. The increased availability of staple foods allowed for specialization. Elite classes of rulers and priests arose. Specialty occupations in ceramics, manufacture of tools, implements, and utensils, and societies for warriors, healers, and leather working became possible.

The increase in food supplies offered by horticulture and gardening provided for a quantum leap in specialties and populations. As efficiencies increased, the resultant refinements in technologies had a "feedback" effect: More efficient tools and implements increased farming efficiencies that allowed for more food, which enabled larger populations, which freed more time for more people to produce more specialty items...and so forth. The Mississippian civilizations are the result of this positive reinforcement between specialties.

Ceramics appear in the Woodlands Phase of the eastern United States as long ago as 2,700 years B.C. Ceramic pottery is comprised of the art of mixing clay with a tempering material, such as bone, fiber, or sand, crushed shell, or even crushed pot shards, then firing the finished piece in an oven. An oven dedicated to the firing of pottery is called a kiln. Caddoan pottery is often bone-tempered, and incised with geometric designs or carved with designs in relief. The pottery found at Caddoan sites can range from rude, sun-baked, untempered pots to highly designed and expertly made fineware.

The advantages of tempering are manifold: By including a material that permeates the clay, heat is transferred throughout the piece with greater efficiency. The pot bakes more evenly and becomes harder. The resulting pot is stronger as a unit and is impervious to liquid. Glazes and slips start out as fine, liquid clay. Glazes include another material element, such as such as mica, copper, lead, or salt, which gives the finished piece a unique color. Glazes and slips can further waterproof the surface by providing a sealant layer. Ceramic containers provided the possibility of storing dried foods over long periods, and for carrying and containing liquids. The efficiencies compounded and translated directly into a healthy, well-fed, and prosperous society.

Artist's concept of a Caddoan village, ca. A.D. 1200. Several dwellings are seen at left; pole beans, maize, and pumpkins are cultivated in the right foreground. In the distance, a ceremonial mound and temple are illustrated. Engraved pottery, bow and arrow, and stone hoes illustrate the Caddoan artifacts discovered during the excavations at the George Davis site in East Texas. Drawing by Jason Eckhardt.

Caddoan Mounds

Prehistoric People in the Eastern Woodlands

The dominant tradition for East Texas has been the Caddoan Mississippian tradition. A versatile and adaptive people, the Caddoans have been agrarians, growing a variety of crops, for more than a thousand years. Caddoans both hunted bison, deer, and small game, and foraged, harvesting naturally occurring fruits, vegetables, nuts, and materials as well. Caddoan trade and diplomacy, together with their populous and well-entrenched cities and villages, gave the Caddoans a very high status among other Indian groups.

The Spanish and French thought so highly of the Caddoan society and its strength, they referred to it as the Tejas Empire, naming the Caddoans after the Caddoan greeting of friendship. Caddoan civilization was so secure, they made no stockades. Due to their long tradition of sedentary life, much is known of the Caddo.

For one thousand years, until their "discovery" by European explorers, the Caddoan people ruled an empire. These were a farming people of the Mississippian culture that occupied parts of present-day Louisiana, Arkansas, and Texas. The Caddo not only participated in but once dominated trade and enterprise in their region, mediating disputes between other tribes, forming trade alliances, and offering protection to friendly Indian bands and punishing enemies. At a time when the majority of Texas Indians were plying an archaic strategy of hunting and gathering or scavenging and harvesting, the Caddo were living in cities that rivaled many in so-called civilized Medieval Europe. They constructed monumental burial and ceremonial earthworks that confounded those who came across them at later times.

The George C. Davis Site — Mysterious Mounds of East Texas

The 14,000 acres that comprise the Davis site contain four major features: two flat-topped ceremonial mounds, a conical burial mound, and a "borrow pit" from which dirt was removed in prehistoric times. The site is bounded on the west by Bowles Creek, which empties into the Neches River to the southwest, about a mile away. The mounds occupy a large, flat terrace, about forty feet higher than the elevation of the river. Approaching the site on State Highway 21, the dense copses of forest are pierced by creeks and punctuated with occasional glimpses of glens and meadows. The atmosphere created by traveling through a tunnel of trees and brush, even at highway speeds, is close and heavy. A sensation of being cloistered and confined is difficult to dispel.

The dense forest gives way immediately upon arrival at the site, contained in the Caddoan Mounds State Historical Site near Alto, Texas. Climbing up and out of the Neches River valley, one of the ceremonial mounds (termed mound A) juts up on the right side of the road. Marked with monuments and interpretive signs, it clearly appears to be worth a stop, but the mound itself, in these days of ubiquitous highway construction, seems no more noteworthy than would a pile of dirt the Highway Department might cache for future use.

In order to correctly assess the impact these mounds had on the first Europeans, it's necessary to imagine weeks of travel, on horseback, through the thickets and dense woods. When the French and then the Spanish explorers came across the mounds, they must have appeared very remarkable. The importance of the area, and of the Caddoan villages that were densely populated, was obvious. The French came to trade, and when the Spanish discovered the French presence and influence, they came to block French

Ceremonial Mound A. Photograph by Robert Marcom, 2001.

Ceremonial Mound B. Photograph by Robert Marcom, 2001.

Ceremonial Mound C. Photograph by Robert Marcom, 2001.

Author's son contemplates Ceremonial Mound C.
Photograph by Robert Marcom, 2001.

ambitions. The Spanish built the first mission in Texas (c. 1690) only a few miles away: Mission San Francisco de los Tejas. The Camino Real (the Spanish Royal Highway) passed through the area and followed ancient Indian trade routes.

The park visitors center and interpretive museum is a few yards further, on the left. A low, unimposing building, it provides a bit of scale for the conical burial mound (called mound C) about 150 yards behind it. Mound B is also a low, unimposing structure located approximately halfway between mound A to the south and mound C to the north. The magnetic north-south alignment, although not obvious to the casual observer, is impossible to ignore when seen plotted on a map. The borrow pit is found in a gully to the west of mound B. The distance from mound A to mound C is approximately 400 yards.

The site has yielded artifacts indicating it has been used for many thousands of years. A Paleoindian camp occupied the southern end of the site, overlooking the Neches River marshes. A larger, Archaic period site—possibly a base camp—overlays the paleo period camp. The site chosen by the prehistoric Caddo for their initial occupation included the earlier sites and extended to the north.

Excavation History

The prehistoric earthworks have been excavated and searched often since their discovery. They've been explored systematically by archaeologists since 1939. The initial work was carried out under the auspices of the Work Progress Administration (WPA). H. Pery Newell supervised the excavations from 1939-1941. He excavated about 60 percent of mound A, uncovering a great deal of trace evidence for Caddoan architecture and building practices in, under, and on top of the mound.

Further excavations were undertaken in 1969. Extensive sampling was conducted at the opposite end, in the northwestern part of the site. About 20 percent of mound B and about 35 percent of mound C was excavated. Surface collections of "virtually all parts of the terrace" (Dee Ann Story, 1998) were conducted. The conclusion was that the site was repeatedly used during all periods of the past: Paleo, Archaic, Woodlands, Prehistoric, and Historic.

Origins of the Caddo

The Caddoan history is divided into periods corresponding to relative dates and radiocarbon dates of artifacts. The culture area of the Caddo includes western Louisiana, northeast Texas, southwest Arkansas, and the Oklahoma side of the eastern reaches of the Red River. The Caddo may have arrived in the area as early as 1,200 years before present (BP).

The geographical zone that includes this site is named Piney Woods. The biome (a biological system that is self-regenerating) has been relatively stable for the period of time covered by the Caddoan occupation. The artifact concentration shows an increase in frequency (more artifacts found) from the Archaic period (6,000 years BP to 2,000 years BP), then a decrease in frequency during the Woodland period, from 2,000 years BP until 1,000 years BP (Timothy K. Perttula, 1998).

Ceramics and the difference in their composition and method of manufacture have been a primary source for dating Caddoan cultural changes. The relative positions of pottery fragments in soil layers yield relative dates. Recently, radiocarbon tests have been available to give absolutes dates within a statistical margin of error. That statistical margin for error has been of much controversy in the past, but laboratory and collection methods have been refined over the past thirty years, until the method is universally accepted as theoretically valid. Individual dates can still be hotly contested, but most recent dates are accepted by most scientists. Dates must be calibrated according to the soil types in which artifacts are found.

The early Caddo are thought to have arrived in the Piney Woods about 1,000 years BP. They probably moved from the coastal prairie. The move was an important one for the prehistoric tribe; it made new resources available. The woods may have inspired larger, more permanent dwellings. If the early Caddo were prairie dwellers, then they probably followed the usual archaic strategy of building low brush-covered shelters. They would have been temporary dwellings, abandoned as the bands moved on in their search for daily sustenance.

There would have been another great advantage to the move—perhaps not immediately obvious to the Caddo. A study of the skeletal biology of prehistoric Caddo (J. C. Rose, M. P. Hoffman, et al, 1998) demonstrates a higher rate of infections and bone disease among Blackland Prairie and of Post Oak zone dwellers than among those who made their homes in the Piney Woods. More than a third of the prairie skeletons showed signs of disease, compared to only a fifth of the pine woods dwellers. Skeletons found in the scrub lands, or Post Oak zone, were intermediate in their rate of disease. There is clear evidence that living in the woods is healthier.

It is easy for those not specifically familiar with archaeological jargon to get lost in the conflicting names proposed for horizons, phases, and foci. For the purpose of this discussion, the Caddoan culture history will be divided into three phases: the Early, also called Formative—A.D. 900 to 1100, Classical, corresponding to the Sanders Phase Caddoan—A.D. 1100 to 1300, and Late, corresponding to the

McCurtain Phase Caddoan—A.D. 1300 to 1700. The Early Phase is differentiated by characteristic pottery and a well-organized social structure indicated by symbolic placement of structures. The earliest phase of mound building may have been for the purpose of elevating the elite above the common residents. The Classical Caddoan period is characterized by fine, engraved pottery, an elaborate system of ceremonial structures, and ritual destruction and renewal of buildings. The Late Caddoan period transitions from prehistory through the Historic period, and it is during this phase that the mounds are abandoned and the Caddoans moved their village to a site a few miles away. Burials occur in both the Early and Classical levels of excavation in mound C, also called the Burial Mound.

Burials

Graves were found on, under, and away from the Burial Mound. The burials indicated elaborate symbolic ritual including preparing the excavated grave with woven cane mats, inclusion of grave goods such as large flint bifaces, stone celts, fine engraved ceramics, and special fill dirt apparently chosen for its color. The oldest burials occurred in the premound stage (Story, 1998), and post holes were found in association with the grave. There may have been a roof over the grave initially, about one meter high, which was mounded over with green clay. The mound collapsed into the grave at some point.

Fewer grave goods were found in earlier burials than in later ones. Grave goods were typically staked against the north wall of the burial pits in the earlier burials, with the exception of the large bifaces and celts. These items evidently identify the deceased as high status persons. The celts, symbolic weapons with an ax-like head and long handle, were made of a variety of green-colored stone and showed some polish and abrasion, as though they were carried in a sheath.

Architecture

The Piney Woods of East Texas offers an abundance of lumber. Oak, hickory, pine, and bois d'arc (bodark) is abundant. Material for thatching is abundant in the marshes, and there is an endless supply of river bottom clay. Buildings could be quite large, some up to twenty meters across. The poles were placed closely in the post holes and were probably thatched over to shed water. Other features included center posts and clay-lined hearths near the center of the structure.

Circular patterns of post holes are found throughout the site, both on the mounds and around them for several hundred meters. The patterns often overlap, indicating that old structures were regularly replaced with newer ones. Burning was sometimes indicated as well. This could have resulted from accidental fires or from ceremonial burning—although ceremonial burning in the pine forest of East Texas seems a risky sort of ritual to this author.

Ceramics

Even the early pottery of the Caddo are well made and decorated with incised lines. Later pottery included engraved designs with functional shapes indicating designs dedicated to specific usage. The types found in centrally located burials included Holly Fine Engraved and Hickory Fine Engraved. Other, nonengraved types found in peripheral burials at the site were identified as Crockett Curvilinear Incised and Davis Incised. Story concludes that engraved pottery was more highly valued than incised and indicates a high status burial.

The Late period of Caddoan culture is documented through the occurrence of French, Spanish, and Texan contacts. The Caddoan Confederacy displayed an organization and sophistication of society that impressed all who came in contact with them. Their agriculture, material goods, and hierarchical government demonstrated a long and evolved civilization. The fact that they persisted over the centuries without need of defensive stockades or barriers gives testimony to the power of their culture.

Summary

At the peak of their influence, the Caddo occupied the fertile stream and river valleys over a vast area including present-day northeast Texas, northwest Arkansas, and northern Louisiana.

The Caddo continue to exist as a tribe, without a dedicated U.S. federal reservation. They have privileges on the Wichita Reservation in Oklahoma (their own reserve having been given in 1874, then taken away in 1901), but many choose to live off-reserve in the vicinity of Anadarko and in Caddo County. The Caddo Nation, once well regarded by Texas republic president Sam Houston, were driven out by pressure from Anglos who coveted their lands and were supported by official policy. The fact that many Caddo sided with the Confederate States during the U.S. Civil War may have cost them sympathy in the post-Civil War federal government. Still, they persist in teaching their language, history, and culture to their children.

Chapter Six

The Spanish and the French—Colonization of Texas

A riddle: Why did the Spanish Empire cross the road? Answer: They thought the French might want whatever was on the other side. The rivalry between these two empires was hotly contested in the New World.

Beginning in the fifteenth century, European powers, including the Netherlands, Spain, Portugal, France, and Great Britain, made concerted forays into newly discovered lands in order to obtain wealth through trade and empire. The Dutch and the English concentrated their main effort on the riches of the Far East, establishing trade colonies by making the long and hazardous voyage around the Cape of South Africa. An Italian noble, Christopher Columbus, had the same objective in mind, but he intended to arrive there by traveling west. Columbus had only one problem; one of which he was blissfully unaware. Columbus had the wrong dimensions for the size of the globe of the Earth.

Columbus correctly understood the shape of the Earth. Greek and, later, Moslem mathematicians had long since divined Earth's shape, and that shape was common knowledge among the educated people of Columbus's day. Columbus thought the Earth was a third smaller than its actual dimension. He placed the shores of India approximately at the location of his landfall in the Caribbean West Indies. Thus the name given to the islands and of the residents, the American Indians. Columbus claimed the new lands in the name of the king and queen of Spain—his sponsors.

The other European powers had no intention of allowing Spain free reign over the New World. The Dutch and English concentrated their colonial efforts in the temperate northeast of the North American continent. The Spanish began their conquest in Central America and became wealthy by plundering the Aztec Empire. They increased their wealth through conquest, moving south to rob the Incas of Peru and Bolivia of their treasures.

The French discovered the main highway of the Mississippi River and its web of navigable tributaries that give access to the interior of the continent. The French began to visualize an opportunity to create a great trade network, channeling the vast resources of North America back to France in return for inexpensive trade items, such as cheap jewelry, utensils, knives, mirrors, and other items the indigenous population had never before seen.

The Spanish fell prey to their lust for gold. They found it perfectly reasonable to believe the tales they were told of more gold possessed by even wealthier Indian empires to the north. They apparently did not suspect that the tales were carefully sculpted lies designed to move the plundering *conquistadores* north and away. The tactic did not work because the Indians misunderstood the goals of Spain.

While the Spanish continued to explore north, they established military forts—called presidios—to protect their new possessions, and religious missions in order to convert the Indians to Christianity. The presidio-mission system would become the signature strategy of Spanish occupation. The military force was intended to protect the Spanish presence, and the missions would provide a

Typical garb of the first European explorers in Texas. The Spanish pike man, ca. 1540, is wearing quilted cotton armor: more commonly employed than the hot, uncomfortable iron armor often depicted. The French musketeer, ca. 1540, is equipped with an early firearm known as a matchlock. Instead of flint striking steel, producing a spark that ignites gunpowder in a pan and setting off the main charge of gunpowder in the barrel behind the ball, a burning cord called a "match" provided the ignition. Drawing by Jason Eckhardt.

Christian workforce to farm and to mine the new lands. Spain was in the Americas to stay.

In 1541 Spanish explorers, under the command of Francisco Vásquez de Coronado, were pursuing the myth of the seven golden cities of Cibola. The Spaniards marched across New Mexico, through the salt flats of the Llano Estacado, across the Texas Panhandle, and possibly as far north and east as western Kansas. They knew there were no riches or great Indian cultures in central and coastal Texas, because a Spanish nobleman named Alvar Nunez Cabeza de Vaca had been shipwrecked in 1528 and stranded in coastal East Texas—probably on Galveston Island near San Luis Pass. He and three other survivors, including his personal slave, Estevanico, made their way to New Spain (Mexico) by 1536 after many adventures among the Indians.

Dominance over the New World lands was hotly contested. Spain was confronted by the English in northern Florida and the Caribbean, by the Portuguese in South America, and in due course, by the French in Texas. The Spanish demonstrated little interest in Texas for the next hundred years. The dense, sometimes impenetrable brush, the materially poor Native American occupants, and the hostile reception of their exploratory probes by many of the tribes encouraged Spanish *conquistadores* to focus their interests elsewhere in the New World. They did not become concerned until, in 1687, they discovered the remains of the French ship *La Belle* in what is now called Matagorda Bay.

Back to the riddle: The longer answer is that the French had no real interest in the other side of the road (Texas); they were lost. The Spanish had no real interest, unless they thought the French did. Given this context, Texas was colonized because of ineptitude and umbrageous envy.

La Salle's Expedition

Born of a misconception of the geography of the New World, Fort Saint Louis was established by Rene Robert La Salle in his search for the Mississippi River. La Salle was certain that the great river would take him to the Gulf of California and from there to the riches of the Orient. La Salle's expedition had vast and unintended consequences that accrued to the benefit of France, but he failed to find the shortcut to riches that he sought.

La Salle's expedition was ill-fated from the beginning. After exploring from the Great Lakes through the midwestern area of the future U.S., he found the Mississippi River. Traveling south on the river, he reached its mouth. Sailing immediately for France, in 1682, he presented his theory to King Louis XIV. After one more trip to Illinois and return to France, King Louis commissioned La Salle to find the mouth of the Mississippi from the Gulf of Mexico. La Salle had a complement of 300 aboard four ships (named *L'Aimable*, *La Belle*, *Le Joly*, and *Saint-Francois*). One ship was seized by pirates before arriving on the Texas coast.

La Salle made landfall at Matagorda Bay on January 1, 1685, but lost another ship and the supplies it carried. A third ship, *Le Joly*, is dispatched to France, leaving La Salle, 180 colonists, and one ship, *La Belle*, behind at some unknown distance from the mouth of the Mississippi River. In April construction began on Fort Saint Louis. Meanwhile, La Salle began searching for the Mississippi by canoe, followed by *La Belle*. The fifty men with La Salle in canoes and the thirty-seven men aboard *La Belle* became separated and had no contact for a month.

La Salle returned to the Matagorda Bay camp and began to search for *La Belle*. A land party finally located the ship, its captain murdered. The captainless ship was wrecked in a storm January 1686. After receiving a report from survivors, who reached Fort Saint Louis after living three months on a peninsula when the ship ran aground, La Salle decided to set out for Illinois on foot. Twenty-two colonists were left behind at Fort Saint Louis; seventeen left with La Salle. Illness and hostile Indians accounted for the fate of the rest of the complement. La Salle himself was assassinated around March 20, 1687.

The Spanish found the wreck of *La Belle* January 1687. They rescued the survivors of Fort Saint Louis the next year, in 1688. The French had been attacked and the Indians had killed the colonists, taking a few children captive. In April 1689 General Alonso de Leon found the remains of Fort Saint Louis. The Spanish responded to that French incursion by building a presidio on top of the ruins of Fort Saint Louis. In order to stave off trade between the Indians and the French, they would undertake to build missions and presidios in East Texas. Although they would be unable to obviate the French claim to the Louisiana territory, they largely succeeded in preventing further French colonization in Texas.

La Belle Project

The wreck of the *Belle* lay in shallow water when the Texas Historical Commission's excavation project began. The problem of how to excavate and yet preserve the maximum amount of context and information was solved by building a wall around the wreck and then pumping the water out. This structure, called a cofferdam, was not a new idea. It had been tried in other places and with mixed results. This time the cofferdam would be constructed in a manner particularly archeological in nature. The design of the cofferdam walls, called bulkheads, would be reminiscent of an ancient construction technique found in classical monumental structures known as "casemate walls."

The cofferdam, octagonal in shape, was constructed in June 1996 with double steel bulkheads, filled in between with sand from the bottom of the bay. Its interior space measured approximately 60 by 80 feet. The cofferdam provided a means to hold back the surrounding bay while the water was pumped from its interior. The last of the water was removed, and excavation began in September 1996.

The excavation was completed in April 1997. Experts have estimated that nearly one million individual artifacts were recovered and catalogued. Of these, approximately 750,000 artifacts were trade beads, colored black, blue, and white. Many lead shot were recovered as well, further

Preparation for the excavation of the wrecked *La Belle* began with the construction of a giant cofferdam around the submerged site. Courtesy of the Texas Historical Commission, Austin, Texas.

The bottom of the bay was exposed by pumping out the seawater. This photo shows the surface prior to excavation. Courtesy of the Texas Historical Commission, Austin, Texas.

swelling the total of artifacts, since each lead shot and bead was cataloged as an individual artifact.

The list of other items found is long. A press release included the following items:

- A human skeleton and additional human bones
- Three ornate bronze cannons with dolphin handles and crests
- A wooden gun carriage, nearly complete
- A swivel gun (small caliber cannon, Ed.)
- Wooden casks containing tools, gunpowder, multiple shot, and pitch
- Wooden boxes containing trade goods and muskets

- Small wooden crucifix with gilded Christ figure and wooden beads, possibly part of a rosary
- Ceramic apothecary jars that once contained medicine, one containing mercury
- Coopers' (barrel makers') plane carpentry tool used to shave down wooden planks
- Pieces of two hourglasses
- Navigational instruments, including eight brass dividers and a compass gimbal
- Two-sided brass pens
- Stacking copper kettles, serving utensils, pewter plates of Sieur Le Gros, a La Salle officer
- French pottery and a majolica Spanish or Italian plate
- Several glass case bottles with threaded pewter caps
- A sewing thimble and a fragment of scissors, probably from a sewing kit
- Articles of clothing, including wooden buttons, fragments of cloth (probably wool), and brass buckles
- A pewter porringer (a shallow bowl with handle) with the name "C. Barange"
- Decorated brass sword guards and pieces of a sword hilt
- Nine leather shoes
- Two complete turtle shells, possibly souvenirs
- A variety of seeds, pits, and other organics
- Two copper flasks
- Bronze candlesticks and a candle dish
- Two 100-foot coils of copper wire
- Bones from deer, dog, bison, pig, bird, and rat
- Cockroach eggs and a variety of insect parts

And more trade items, including:

- Brass Jesuit rings
- Hawk bells

- Brass straight pins
- Combs and mirrors
- Red ocher, probably for use as body decoration (by Indians, Ed.)

Additionally, ship rigging and equipment including deadeyes, cleats, 600 feet of anchor hawser, a bilge pump complete with much of its leather pump apparatus, and four brass sail makers needles.

There are plans for a major permanent exhibit in a Texas museum, yet to be selected. Many of the artifacts were displayed in Victoria, Texas, at the Texas Historical Commission's Public Archaeology Laboratory during

This photo illustrates the careful, systematic process that ultimately yielded even those fragile cloth, paper, and leather artifacts, as well as two ornate bronze cannons, utensils, and a human skeleton. Courtesy of the Texas Historical Commission, Austin, Texas.

Archeology Month, 2001. Plans to tour the exhibit include statewide, national, and international exhibitions. Meanwhile, restoration and preservation work continues, as well as work on interpretation. Private donations, grants, and gifts are being sought by THC in order to finance the work and tours.

The fine preservation of these artifacts, including bone, cloth, and leather items, which normally deteriorate rapidly, is due to the fact that the wreck was sealed in mud. The number and condition of the artifacts has been called amazing. The human skull contains brain tissue, and the skeletal remains still have connecting tissue attached. Researchers completed a reconstruction of the face of the sailor. The skeleton was discovered in October 1996. A three-dimensional image was created by technicians at the Scottish Rite Hospital for Children in Dallas, then an exact duplicate of the skull was created by CyberForm of Richardson, Texas. Professor Denis Lee at the University of Michigan constructed a clay model of the face, created a silicon mold, then reconstructed the facial features. A press release by THC, October 27, 1997 says of the face:

The face of the sailor whose skeleton was found in the wreckage of *La Belle*. Features reconstructed from the skull by CyberForm International, Richardson Texas. Photo by Robert Marcom, 2001.

"While it may be possible to 'put a face on history,' scientists may never be able to determine the exact cause of the sailor's death. It is believed he was of European descent, had a stocky build, a broken nose and bad dental cavities. He may have died from dehydration, hypothermia or drowning." The press release continues, "A pewter porringer engraved with the French surname 'Barange' found near the skeleton could be the missing link to the man's identity."

One mystery may remain forever. Inexplicably, a 2,000-year-old Roman coin, a dinarius dating to A.D. 69, was discovered among the artifacts. It was the only coin found in the wreck.

This ornate bronze cannon is one of two found below decks on the *Belle*. This cannon bears two crests: the crest of Louis XIV near the breach and the crest of Le Comte de Vermandoise, Admiral of France, near the center. Photograph by Robert Marcom, 2002, Courtesy of the Public Archeology Laboratory, Victoria, Texas.

This scale model of the 17th-century French ship *La Belle* was on display at the THC Public Archeology Laboratory, Victoria Texas. Photograph by Robert Marcom, 2002, Courtesy of the Public Archeology Laboratory, Victoria,

Fort Saint Louis

The Archaeology Division of the Texas Historical Commission sponsored the excavation of the fort, which resulted in the recovery of tens of thousands of artifacts, including human remains, a cache of eight cannons, French pottery, personal effects, and the footing trench of the original walls of the fort. The French fort constitutes the earliest European occupation in Texas.

The main building of the French fort was constructed of timbers salvaged from one of the two wrecked ships. The human remains of two, and possibly three, French colonists confirms the account given by the Spanish of finding the bodies and burying them.

The remains of the original Spanish presidio, built on the ruins of Fort Saint Louis were found, along with a great number of Spanish artifacts. The elaborate Spanish fortification was constructed of logs set upright in the shape of a sixteen-point star.

Presidio La Bahía Nuestra Señora de Loreto, at Goliad, Texas

Presently located a few miles south of Goliad, Texas, the presidio, *Nuestra Señora Santa María de Loreto de la Bahía del Espíritu Santo,* was originally built on the ruins of the French Fort Saint Louis. The Spanish called the presidio "La Bahia" for short. The military post was established to prevent French claims in the area of the bay, which the Spanish named *La Bahia Espiritu Santo*, or the Bay of the Holy Spirit.

The presidio Bahia was relocated at least twice from the original location, further inland. From its establishment, the presidio in its current location became the most important destination in Texas on the Camino Real, the Spanish Royal Highway. The presidio became protector for the mission *Nuestra Señora del Espiritu Santo del Zuniga* and the focus of a community of Spanish, Indian, and mestizo residents.

The Camino Real stretched from Mexico to the Spanish colony at Nacogdoches and intersected with the Atascocito Road near La Bahia, connecting the community, presidio, and mission with San Antonio. The name of the community was changed by petition to the Spanish authorities in 1726, in response to the argument that the name was meaningless, since neither the presidio, mission, nor the villa (town) was located by the bay. The legislature of the state of Coahuila y Texas renamed the town Goliad and conferred the official status of Villa on the community.

La Bahia Excavation

During the final week of Texas Archaeology Awareness Month, October 2001, I visited Presidio La Bahia for the culmination of an excavation open to the general public. Three units were opened under the supervision of Mr. Jeff Durst, Texas Historical Commission Archeologist for Region 6, and Mr. Newton Warzecka, Director of the Presidio La Bahia, Diocese of the Victoria Catholic Church. On the day we visited, Ms. Anne Fox, Laboratory Director for Presidio La Bahia, was unit supervisor.

Unit 1 was thought to be located on a trash midden. The purpose, beyond demonstration of excavation techniques, was to confirm the trash pile location. Unit 1 was expanded, as an artifact in the wall of the unit indicated they might be on the edge of the trash midden.

The volunteers, digging in the units and passing the dirt through screens as it was removed, included a broad spectrum of people, ranging from professional and avocational

Presidio La Bahia excavation demonstration for Archeology Awareness Month, 2001 Left: shaded units being excavated. Center: archeologists surveying the site. Photograph by Robert Marcom, 2001.

Presidio La Bahia, Goliad Texas. The chapel (left) looms above the fortress walls. Photograph by Robert Marcom, 2001.

archaeologists, to interested laypeople. The work was carefully supervised and the results documented. The dig took place over the weekend of October 19-20.

The units were located by the use of an electronic surveying instrument, called a Total Data Station, operated by Mr. Bill Pierson (assisted by Jeff Durst). The electronic transit was complimented by the use of a stadia rod with a built-in laser target. The angle and distance of the rod's target was recorded and calculations were recorded.

On the day I visited the excavation, I noted a camaraderie among the digging crew that I've often seen at digs. The intellectual exercise has a strong physical component: Artifacts and their associations are only found after dirt is removed and screened. Conversation helps to pass the time and develop a rhythm. A tone of easy competition inevitably arises, and did so here as the dig progressed. The individuals at each unit became a group, and the groups became a team.

The warm fall day and the notably fine weather contributed to an atmosphere I can only describe as picnic-like. Visitors drifted through in family groups, stopped, asked questions, and made comments. The archaeologists discussed features, such as discoloration of soils, the occasional pottery fragment, or the ubiquitous nuisance of tree roots. An easy ambiance was punctuated by chatter, banter, dripping sweat, and mild exclamation whenever an item of interest surfaced in the unit or the screen.

Results of the first unit included pottery and animal bones (possibly dietary refuse). The pottery types were tentatively identified as English ware, transfer ware, and possible Rockport ware. The site of a trash pile does seem to be indicated.

Two more units were opened to investigate metal detector "hits." A unit near the center of the enclosed grassy area produced a fine example of a brass rifle lock, the portion of the weapon that contains the trigger, cock (hammer, which

Dirt excavated during the Presidio La Bahia demonstration is carefully screened. Artifacts are bagged and provenience information recorded. Photograph by Robert Marcom, 2001.

holds a rifle flint) pan, and strike plate. The third unit, near a wall of the fort produced nothing of note.

A tour of the museum, located in the rooms that had been administrative offices, offers a fine array of exhibits, including artifacts, re-creation of an oven, and historical interpretation. La Bahia served the Spanish Empire, the Mexican Republic, and the Republic of Texas. The presidio, church, and fortress walls have been meticulously restored. The ramparts and blockhouses are equipped with cannon dating from the Republic of Texas, Fort Defiance historical period.

Nearby, the Texas State Historical Park, Mission Espiritu Santo, offers a comprehensive interpretation of the history of the area with dioramas, artifacts, and models.

Chapter Seven

Tejas and the "Gringo" Invasion

The province of Texas was administered by Spain until 1827. Except for the areas around San Antonio de Bexar and the few Spanish missions established in East Texas, the Spanish governors regarded the province as an unproductive and largely impassible land, occupied by hostile tribes of intractable Indians. Mission after mission was attempted alongside the rivers of the Texas coastal plain, and one after the other they were abandoned. The local Indians would not stay "converted" to Christianity, and the missions were constantly under threat of being raided by Apaches, Comanches, and Kiowa. The most successful Spanish missions were those that had a strong military contingent.

The Spanish administrative province of Texas included present-day Texas and parts of New Mexico and Colorado. Spain adopted a policy of encouraging settlement of Texas by immigrants from the United States. The immigrants agreed to become Mexican citizens and, in return, were given grants of land. The best known of these immigrants was Stephen F. Austin, who arrived in 1824. Austin's father, Moses Austin, received a grant of land from Spain in 1821. Stephen F. Austin's colony was comprised of the "Old Three Hundred," the first U.S. citizens to settle on land granted to Moses Austin.

The Texas Forts

Texas became an independent republic after the defeat of Mexico's army, led by General Antonio Lopez de Santa Anna, at the Battle of San Jacinto in 1836. Texans soon realized they would need a means for defense against hostile Indians. The Texans and the Lipan Apaches allied against the Comanche, who were traditional enemies of the Lipan.

The Comanches arrived for talks in San Antonio de Bexar in 1840. They were to discuss a treaty and the exchange of hostages held by the Comanche from previous skirmishes. The talks broke down when the Texans demanded the release of all white prisoners and the removal of all Comanches from central Texas. The Texans informed the Comanche chiefs that they were hostages. A battle broke out, called the Council House Fight, in which several Indian chiefs, women, and children were killed in an effort to escape the vicinity of the Council House.

The Comanches were outraged at the act of war made on them while they were in council. From their perspective, they could not comply with the demands because they had no authority over the Comanche tribes not attending the council. The Comanches regarded the act of the Texans as treacherous, and they began raiding deep into Texas.

For two years, through efforts made by President Sam Houston and others, the Apaches assisted the Texans in the effort to stave off the Comanche onslaught. The Lipan Apache band enlisted and served with the fledgling Texas Rangers in campaigns against hostile Indians and against the Mexican raiders who constantly probed the border lands.

In 1839 Texas president Mirabeau Buonaparte Lamar vowed to rid Texas of all Indians. The Lipan may have had a sense of foreboding, with the expulsion of the Cherokees by

Texas in 1839 and the Council House Fight in 1840, but they continued to assist Texas in its skirmishes with Mexican troops and hostile Indians until Texas Rangers attacked a Lipan village near present-day Castroville in 1842, as punishment for supposed Lipan raids against settlers. The Apaches fled Texas and began their own cross-border assault on Texas.

Texans had larger worries in the form of a determination by Mexico to recapture the territory lost in their 1836 defeat. Skirmishes escalated to large-scale battles, including the Battle at Salado Creek. In September 1842 Texas Ranger Colonel Mathew Caldwell and Texas Ranger Captain John "Jack" Coffee Hayes led a group of 600 Texas Rangers and volunteers that defeated a Mexican force of 1,400 soldiers north of San Antonio.

Captain Hayes turned his attention toward the Comanches in 1844 and decisively defeated them at the Battle of Walker Creek. The Texas Rangers employed the Colt Patterson 5-shot revolver for the first time at Walker Creek. A Comanche chief was quoted after the battle as having said, "I will not fight Walker for he has a bullet for every finger of his hand."

Sam Houston, past president and future governor of Texas, engaged in political maneuvers to achieve the annexation of Texas by the United States. The U.S. Congress rejected annexation in 1844 for fear of entanglement in the Texans' conflict with Mexico and over questions regarding the balance of slave versus free states and territories in the U.S. The United States Congress passed the annexation resolution on February 28, 1845. The U.S. Army did go to war with Mexico. The Texas Rangers participated as scouts and guerilla fighters. After the war, the U.S. Army established new forts and assumed authority over existing forts.

Republic of Texas Forts

The Alamo

The stirring rally cry of "Remember Goliad! Remember the Alamo!" has rung down through the last 170-plus years as a testimony to the valor of a few volunteers—both Texans and others—who stood against the thousands of soldiers of General Antonio Lopez de Santa Anna. The battle has been recounted many times and in many ways: in articles, essays, books, movies, and television dramas. The story is well known but continues to intrigue and stimulate wonder:

"Thermopylae had its messenger of defeat. The Alamo had none."

"The Alamo in 1836 consisted of this church, the convent and a large rectangular area or plaza, an enclosure of about six acres surrounded by walls with barracks on the west side of the Plaza. On February 23, 1836 Colonel William Barret Travis entered the Alamo with an approximate force of two hundred men. The siege, commanded by General Santa Anna and an army of several thousand Mexican soldiers lasted nearly two weeks. At dawn on Sunday, March 6, the final assault was made, and in less than an hour the defenders slain. Later, the bodies were burned by order of General Santa Anna.

"This victory in defeat was the means of uniting the colonists in a determined effort to resist further oppression and by armed force to secure permanent independence.

"It was here that a gallant few, the bravest of the brave, threw themselves between the enemy and the settlements, determined never to surrender nor retreat. They redeemed their pledge to Texas with the forfeit of their lives. They fell the chosen sacrifice to Texas freedom." (Historical Marker, Alamo Plaza, San Antonio, Texas)

Whether that sacrifice was necessary or ill advised and against orders is a matter for historians, but the valor of deeds performed leaves little room for critique. The building they fortified was built to house the Mission San Antonio de Valero. The mission was authorized by Spain in 1716. It was moved to its present site in 1724, and in 1803 it was first used as a military fort. The Alamo changed hands twice during the Mexican Revolution, being lost and retaken by Spain. By 1824 the Alamo was firmly in the hands of Mexico. The Mexican administrative state of Coahuila y Texas was formed, with the state capital at Saltillo, Mexico.

The buildings and grounds were purchased with an appropriation by the Texas Legislature in 1905 and delivered "into the custody and care of the Daughters of the Republic of Texas." The expanded battleground was purchased, and in 1960, the Alamo was designated a National Historic Landmark. The grounds house a museum with an extensive collection of artifacts, displays, and interpretive plaques.

Camp Salado

Established in 1842 after the Battle of Salado Creek, this fort is six miles northwest of San Antonio, Texas. It is currently within the bounds of U.S. Army post Fort Sam Houston, the home of Headquarters, U.S. Army Medical Command. It was, in the days of the republic, a Texas Rangers outpost. Its history begins as a campground used by Colonel Mathew Caldwell for the assembly of the volunteers and rangers who would fight against the Mexican army led by General Adrian Woll. The Mexican army recaptured San Antonio in September 1842. Caldwell called for volunteers to repel the Mexican army, and Captain Nicholas M. Dawson responded with a company of fifty-three volunteers, marching down from La Grange.

Dawson's company ran into trouble before reaching Caldwell's camp. He encountered 500 Mexican cavalry on the grounds of present-day U.S. Fort Sam Houston. Dawson's company fought off a cavalry charge. The Mexicans withdrew out of rifle range then began to shell Dawson's men with artillery. Dawson made an attempt to surrender, but the Mexicans shot him down, according to survivors of the encounter. Thirty-six Texans died, fifteen were captured, and two escaped. The "Dawson Massacre" became a rallying event, and the Mexicans were forced to retreat toward the Rio Grande River two days later.

Fort Parker

In 1834 Silas M. and James W. Parker built a private fort near present-day Groesbeck in Limestone County, Texas. The Parker family brought the Predestination Baptist Church membership to Texas. Near the headwaters of the Navasota River, they built a wood stockade and cabins to shelter the church members in case of an Indian attack. On May 19, 1836, Comanche and Kiowa Indians (Comanche and Caddo, by some accounts) attacked, killing many of the occupants and taking some of the women and children. Among the captives was young Cynthia Ann Parker. Those settlers who lived through the attack fled to Fort Houston, near Palestine, for safety.

Cynthia Ann Parker grew up as a Comanche, living with the Tenewa and Yamparika bands and marrying Peta Nocona. She had three children, two boys and a girl. One of her sons would grow up to become the Comanche war chief Quanah Parker. Her other son was named Pecos and her daughter Topsannah.

Cynthia Ann Parker was captured by the Texas Rangers in 1860 during a raid on a Comanche camp on Mule Creek, a tributary of the Pease River. Cynthia Ann did not return to life among the whites willingly, and she considered herself Comanche until her death.

A replica of Fort Parker (perhaps named Fort Sterling, originally) is located on State Highway 164, northwest of Groesbeck. The original burial place of the victims of the Indian raid is nearby, in Memorial Park Cemetery, on State Highway 14.

Fort Houston

This fort was established in 1835-36 near Palestine, Texas, in order to support settlers in East Texas. The fort was built by the Republic of Texas and consisted of a blockhouse and stockade. It was the site of one of the earliest Texas Ranger units. A historical marker at the junction of U.S. Highway 79 and FM 1990 commemorates the fort.

U.S. Army Forts

Fort Sam Houston

Established in 1845, Fort Sam Houston is home to Headquarters, U.S. Army Medical Command. The Museum Division operates a museum open to the public. Admission is free. Tours and historic interpretation, as well as displays of artifacts, are available Wednesday through Sunday, 10 a.m. to 4 p.m. The museum is 2½ miles from San Antonio at 1207 Stanley Road, Building 123, Fort Sam Houston, Texas 78234.

The fort is the ninth oldest active post in the U.S. It was once host to an experimental program to explore the use of camels by the army. According to museum information, the fort has the largest collection of historic buildings of any active military installation—and nine times as many as colonial Williamsburg, Va.

Texas Forts Trail

The Texas Historical Commission sponsors a tour of U.S. forts established beginning in 1848. The Texas Forts Trail literature says, "The 650-mile trail covers 29 counties in West Central Texas and highlights eight historic frontier forts and the communities and attractions that surround them. The remains of these eight frontier forts offer a tangible link to an early period of Texas history. Featured are Fort Belknap near Graham; Fort Chadbourne near Bronte; Fort Concho in San Angelo; Fort Griffin north of Albany; Fort Mason in Mason; Fort McKavett outside Menard; Fort Phantom Hill near Abilene; and Fort Richardson near Jacksboro. The site of a presidio near Menard that dates to the Spanish Colonial period is part of the trail as well."

Fort McKavett — An Authentic Frontier Fort

Fifty miles south of the modest but modern city called San Angelo, Texas, and more than one hundred twenty years ago, a collection of a few hundred soldiers stood guard at the far frontier of a young and growing nation. I had the pleasure of receiving a guided tour of the fort those stalwart defenders called home.

Fort McKavett State Historical Park is comprised of the preserved buildings, artifacts, and spirit of the fifty-three or more companies of infantry and cavalry that manned the ragged edges of a new country. The historical park contains the only completely original buildings of the chain of Texas frontier forts.

The visitors center is located in the buildings that housed the Post Hospital. The diorama opposite the admission counter is well worth a moment of contemplation. It will disabuse one of the images of frontier army life that has been provided to us by movies and television.

This is a fort without walls or stockades. Fort McKavett occupies the highest open ground in the area. The soldiers routinely patrolled an area of more than one hundred miles in diameter. Contrary to the image of an island of besieged troops standing off hoards of Native American attackers, Fort McKavett's complement dominated the area with confidence.

Leaving the visitors center, walking out through the shaded "dog run" between the hospital buildings, the view is one of contrasting ruins and meticulously restored buildings. The flagpole is visible across the entire grounds and serves as a prominent reference. The flagpole is a sixty-two-foot-tall pine mast from which a large American flag flies. It is set in the midst of a large, open parade ground and provides both a landmark and a focus for the collection of quarters for officers and enlisted men, the headquarters building where daily reports were no doubt ground out in

View of the parade ground, Fort McKavett. Photograph by Robert Marcom, 1999.

Ruin of the Headquarters Building of Fort McKavett. Photograph by Robert Marcom, 1999.

triplicate, and the large, foreboding ruins of the Commanding Officers Quarters building, which seems to hover over the extensive collection of buildings and ruins.

Fort McKavett is a repository for the traditions of a hallmark event of American history. Early attempts at racial accommodation in the post-Civil War era led to the creation of African American units, which were christened the "Buffalo Soldiers" by Native Americans. Fort McKavett was home to several regiments of black infantry and cavalry.

Park Superintendent Buddy Garza and his staff are well versed in the history of the site and readily provide historical interpretation of the lives and times of the men who served there. Superintendent Garza has served as a Texas Parks and Wildlife ranger at the fort for several years and has been in charge of the park for the last three years. His philosophy is simple: "We want to provide the authentic experience of frontier life—the natural feel of a frontier fort." Ranger Garza went on to explain that Fort McKavett is the only Texas frontier fort composed mainly of original restorations—not replica buildings.

As one visits each building, it is quite easy to conjure images of the past. Many of the buildings are furnished with authentic, original, and replica items. The hard, straw-stuffed mattresses of the barracks, the fire-blackened stones of the fireplace hearth, the neatly folded gray blankets with "US" emblazoned, hint at the spartan lives of the privates who once protected the prairies. Headquarters building, with its cast-iron stove and plain green furniture, brings images of a grizzled first sergeant receiving the morning report from the regimental NCOs.

Ranger Ed Quiroz provided a demonstration of the infantryman's uniform and weaponry. He also debunked some common mythology regarding the military of the period.

"The post-Civil War period is one of the most misunderstood periods in American history. They didn't wear yellow bandanas and they didn't have stripes on their pant legs, unless they were officers. Officers didn't wear white hats."

He went on to explain that the cavalry played a much smaller role, and the infantry a much greater one, than you would suspect from TV and movies' portrayal of the times. That, he said, is the reason he chose to interpret the life of an infantry private rather than the more glorified cavalry trooper.

Standing on the wind-swept parade ground, the sun casting long fingers of shadow across it, one may capture a sense of the isolation and loneliness many soldiers must have felt during their service. The post was one of honor and opportunity for officers seeking combat experience for their resumes, but it must have been a difficult and wearing time for the enlisted men. Texans can be grateful to the few hundred who served at Fort McKavett during a turbulent time in the history of our country.

Many of the restored buildings are furnished, and there are displayed artifacts and interpretive signs. Visitors are subject to the customary state park rules and use fees.

Typical uniform and kit for a U.S. Army enlisted man, ca 1860. Demonstrated by Texas Park Ranger Ed Quiroz. Photograph by Robert Marcom, 1999.

Chapter Eight

The Old Frontier: The Red River Wars

During the period from 1868 to 1875, the panhandle areas of present-day Oklahoma and Texas were the scenes of raids by Native American Southern Plains tribes including the Arapaho, Comanche, Southern Cheyenne, Kiowa, and Kiowa-Apache. These tribes were attempting to defend against the Anglo-Americans' invasions of their traditional hunting grounds and their winter quarters.

In 1867 the U.S. Congress created an Indian Peace Commission. The Commission held a conference with the Southern Plains tribes near Fort Larned, Kansas. Under threat and ultimatum, the tribes were forced to agree to presents and annuities in return for relocating to reserves in the Indian Territories, present-day Oklahoma. The treaty guaranteed, in addition to annuities and gifts, the right of the Indians to hunt on any lands south of the Arkansas River "so long as the buffalo may range thereon." When U.S. government provisions proved to be insufficient to sustain the tribes, they returned to their hunting grounds.

The Indians saw their last remaining herds of buffalo relentlessly slaughtered on the Kansas plains. Railroads, the U.S. Army, and entrepreneurs destroyed the northern herds then began to hunt south of the Arkansas. General Philip Sheridan, whose duty it was to prevent Anglo incursion into the Indian hunting grounds, instead urged the extermination of the herds. In 1874 buffalo hunters moved into the area of the Canadian River from Dodge City, Kansas. A trading post was established at Adobe Walls to support the large group of hunters.

This sign is near the original location of the hunting camp. Note the trees, indicating a nearby source of water. Both Native Americans and bison hunters required a source for water and wood. This basic requirement usually brought conflict. Photograph by Robert Marcom, 2002.

By 1874 Native Americans may have felt the inevitability of the end of their way of life, but they were not ready to surrender that lifestyle willingly. On several occasions, raids by independent groups of warriors resulted in rapes, beatings, and murders of farmers and settlers within the south plains region. This region was bounded on the north by the Arkansas River (southern Colorado) and on the south by the Salt Fork of the Red River (Panhandle, Texas). Eleven major engagements and numerous minor engagements and skirmishes finally forced the tribes onto reservations, disarmed them, and made them fully dependent on U.S. federal programs for food and shelter.

Many tales were told of the suffering, hardships, and depredations suffered by the Anglo-Americans during this struggle for eminence. Some of the fondest-held myths and legends of the American Old West are derived from the historical record of these events. Much of what everyone "knows" about the Plains Indian, the U.S. Calvary, the buffalo hunter, and life on the western plains is now being reexamined by archaeologists through a programmed excavation of battle sites.

The Panhandle Plains Historical Museum's curator of archaeology, Dr. Jeff Indeck, said of the two battle sites "The two battles at 'Adobe Walls' are distinguished archaeologically by the fact that the sites are about one mile apart. The first battle, with Kit Carson, occurred on November 26, 1864, near the site of an old William Bent trading post that was constructed about 1843. This original structure consisted of a fort constructed with adobe. The post was abandoned due to hostilities with Southern Plains Indians, but the structure became a well-known landmark. Although supplied with two mountain howitzers, Carson eventually retreated from the area due to the great numbers of Indians in the vicinity."

The Red River War commenced with the second battle at Adobe Walls, on the south fork of the Canadian River near present-day Borger, Texas. The Panhandle area was the scene of numerous engagements and clashes between the Cheyenne, Comanche, and Kiowa and the U.S. Army for more than a decade. Such notables as William F. "Buffalo Bill" Cody, William "Wild Bill" Hickok, and Bat Masterson scouted and fought in the region for the U.S. Army. General (brevet) George Armstrong Custer made a foray in strength into the area at the head of the 7th U.S. Cavalry in 1868-69. He barely escaped with his life, presaging his last stand five years later at the Battle of the Little Bighorn. In June of 1874 the control of the winter campgrounds was by no means guaranteed to the Anglo-Americans.

Second Battle of Adobe Walls

Adobe Walls sits astride the north-south migratory route for the great southern bison herd. At the time of the Second Battle of Adobe Walls, the settlement was comprised of four business concerns: the Rath and Wright hides store and corral, the Hanrahan saloon, O'Keef's blacksmith shop, and Myer's and Leonard's general store.

The site was named for an adobe fence, long-since melted away. The scarcity of building materials and the great expense entailed in transporting lumber dictated that these buildings would likewise be constructed out of adobe (mud and straw) bricks. There were numerous campsites and dugouts surrounding the settlement. The area is at the transition between the flat, open prairie and the "breaks" of the Canadian River. As the ground descends to the floor of the river valley, it is eroded. The erosion creates narrow, steep arroyos and gulches known as "breaks."

In the eight years following the end of the Civil War, the U.S. Army turned its attention to the protection of U.S. citizens on the frontier. Forts and depots were established in Texas in a string from the Rio Grande Valley to Ft. Stockton in West Texas. Anglo-Americans pushed out beyond the line of forts wherever they found economic incentive to do so. The Adobe Walls settlement was a business venture designed to take advantage of the demand for buffalo hides. Native American tribes saw the Anglo presence as an unwarranted intrusion on their traditional way of life. By

1874 their experiences with Anglos brought them to believe there could not be a peaceful coexistence. On June 24, 1874, a combined force of Cheyenne, Comanche, Arapaho, and Kiowa warriors besieged the settlement at Adobe Walls. The battle lasted for three days and resulted in four Anglo and an unknown number of Native American deaths.

The battlegrounds of the Second Battle of Adobe Walls. Photograph by Robert Marcom, 2002.

Nothing remains of the settlement today. Only a marker placed by the Panhandle-Plains Historical Society indicates the site of one of the most ferociously fought engagements in the history of the Old West. Perhaps the most stirring account of the battle is found in the recounting of the battle by Billy Dixon to his wife, Olive K. Dixon, as recorded by Bob Izzard in his book, *Adobe Walls Wars*. Izzard's book recounts harrowing tales of desperate runs between the various buildings in order to acquire and distribute ammunition and buoy the spirits of the fighters. The banter between notables such as Billy Ogg, Dixon, Bat Masterson, the Hanrahans, and others rings true to an ear tuned to the vernacular of the times. Within the events and stunts of the engagement can be found much of the stock and boilerplate that has given plots to innumerable horse operas for stage, screen, and television.

Oral reports of the battle are notably one-sided. According to the Anglo-Americans present at the battle, a force of 700 to 1,000 warriors attacked the settlement at dawn. The assault seemed to be organized to some degree, with warriors utilizing bugle calls and coordinated assaults from different directions. Native Americans fired arrows and firearms from cover and used terrain features as firing positions. The settlement defenders were well armed and possessed among them some very fine marksmen. This would be easy to believe, since buffalo hunters typically hunt from a distance, killing bison one at a time in order to keep the herd from stampeding. The term for this situation is called a "stand." Often, hundreds of buffalo could be killed over a period of hours.

The Adobe Walls battle site is now in private hands. The owner has allowed excavation and the placement of markers commemorating both battles and accommodating the burial site of one of the heroes of Adobe Walls, Billy Dixon. Dixon was the first of the hunters to see the warriors and was instrumental in the defense of the settlement. Dixon later won the Congressional Medal of Honor for his heroic performance at the battle of Buffalo Wallow during the Red River Campaign.

Artifacts recovered from the site include spent munitions and the normal refuse one would expect for the period such as leaded tin cans from rations, horseshoes and nails, as well as miscellaneous broken tack. The pattern of spent

The grave marker for Billy Dixon, Medal of Honor recipient and combatant at the Second Battle of Adobe Walls. Photograph by Robert Marcom, 2002.

cartridges as well as the unique characteristics they share indicate there were fewer than the reported 700-1,000 warriors in the raiding party. Rifling (the grooves and lands carved into bullets from the barrel of firearms) and unique marks left by the firing pins and extractor mechanisms on the shell casings made possible the identification of many of the weapons used during the individual battles. Archaeological results tend to support the Indian version of the numbers involved—possibly 300 to 400 warriors participated under four different war chiefs. As is usual with Plains Indian warriors, they were free to fight or to leave at will.

The popular version of the battle includes several elements that have been discussed and disputed by the occupants of Adobe Walls. Most but not all accounts credit

Monument to the Native Americans who fought at the Second Battle of Adobe Walls. Photograph by Robert Marcom, 2002.

the alertness of the buffalo hunters to the cracking of a ridge pole holding up the roof of Hanrahan's saloon. The noise purportedly awakened several of the hunters sleeping in the establishment, including Billy Dixon, who is credited with raising the alarm. There is no evidence of Dixon having made this claim, but it has often been recounted by historians. Izzard includes the event in his recitation of recollections by Dixon's widow.

Izzard also includes an Indian who blew bugle calls (a black man in some accounts), and a 1,538-yard-long shot, made by Billy Dixon, that seemingly discouraged the remaining Indians from continuing the assault and siege. Of these details, Dr. Indeck comments, "T. Lindsay Baker and Billy Harrison summarize the archaeological evidence in the book *Adobe Walls: The history and archeology of the*

Monument to the buffalo hunters and frontiersmen who fought at the Second Battle of Adobe Wells. Photograph by Robert Marcom, 2002.

1874 trading post. I believe that one of the most difficult aspects of the authors' research was that many of the accounts were recorded by individuals other than participants, and they were documented significantly after the event. Consequently, many of the recollections are inconsistent. There is great debate about the veracity of the 'Billy Dixon long shot,' which still receives articles written about it, as recently as last month. Two 'facts' are worth considering, however. There is no hilltop 1,538 yards away from the structures, and Dixon never made claims about the shot in public. It is only through a widow's biography that the details are provided, and then, forty years after the fact. It is also interesting that details about the shot change between the first and second editions of the biography.

"There is no archaeological evidence available that would shed light on questions about who the participants were, whether there was a black bugler, whether the saloon roof beam cracked, how many Indians attacked, or how many Indians were killed. As Baker and Harrison provide, researchers are directed to the information and they may then form their own opinions. The archaeological evidence does provide information about the size, construction materials, and locations of the buildings; structure furnishings; tools, equipment, and personal items; and objects for sale or trade. A few interesting activities were identified, including a blacksmith who kept his beer cool in his quench tank, hunters eating with china plates and cups, and glass windows that had to be carried in wagons from Dodge City, Kansas."

The likely participants on the Indian side included about 300 Comanche warriors of the Quahadi band led by Isa-tai and Quanah Parker, a few dozen Kiowa warriors under the leadership of Lone Wolf, and the remainder composed of Arapaho and Southern Cheyenne Indians. The attack failed dramatically due to the alertness of the buffalo hunters in the Adobe Walls compound and to the fact that buffalo hunters were excellent marksmen, firing from well-protected positions, and the fact that the general store had a large stock of ammunition of all calibers. Two hunters and one of the storeowners were killed in the raid, and numerous Indians died.

The result of the June 1874 battle at Adobe Walls was reported to the army at Fort Dodge, Kansas. By August the army responded. Five columns of troops were dispatched from forts surrounding the Red River War battle sites. Two columns were to approach from the south and west, from Fort Richardson and Fort Concho, Texas. Two were sent west and southwest, from Fort Sill and Camp Supply Oklahoma (then known as the Indian Territories). One column was to move east from Fort Union, New Mexico. The column from Camp Supply engaged the Indians in the first major battle.

Battle of Red River

The Red River battle took place on August 30 in the "breaks" of the Red River (now called the Prairie Dog Town Fork) near Palo Duro Canyon, Texas. The site is privately owned and not available to visitors. In 1998 the Archeology Division of the Texas Historical Commission surveyed the area traditionally understood to be the battle site as part of the Red River War Battle Sites project.

Walking near the area, in similar breaks, I imagined the difficulty of maneuver that must have confronted the soldiers. The land is deeply eroded by the infrequent but torrential rains as they run off into the riverbed. The soil is a loose sand of red iron oxide called Permian Red Bed. Scrambling up and down the gullies and arroyos, I imagined what it must have been like for the soldiers, trying to move cannon and Gatling guns into position, attempting to maintain contact with supporting units, under fire by Indians who could pick the time and place of engagement.

The army had available one 10-pound Parrott rifle (small cannon), two 10-barreled Gatling guns, and 744 men organized into two battalions of cavalry, infantry, and artillery. The Indians opposing them were thought to number 400 to 600 warriors, mostly Southern Cheyenne, with a few Kiowa and Comanche. The battle lasted for the better part of two days, as the Indians fought a rear-guard action in order to allow their families to escape across the desert wilderness of the Llano Estacado to the west, and to safety. The combat took place on the first day. According to report, the second day was spent in anticipation the Indians would regroup.

They did not, preferring to join their fleeing families. One soldier was killed and the Indians lost about twenty-five warriors.

Early accounts of the battle and local lore led to the assumption that the army had proceeded through Wagon Wheel Gap to the west of Battle Creek—a small tributary of the Prairie Dog Town Fork. The assumption appears to make sense on its surface in light of the amount of equipment transported and the better footing and travel conditions for horses. This assumption proved to be incorrect. The distribution of munitions leaves no doubt that the battle unfolded on less favorable ground to the east and southeast.

The artifact distribution indicates the battle commenced on the north side of Battle Creek at Griffin's Hills. Artillery barrage is evidenced by the recovery of Parrott gun shell pieces south and west of a prominence called Cannon Hill, a name that reflects its use during the battle. Hardware and tack, along with spent ordnance, tells a tale of the vastly outgunned warriors fighting desperately and successfully to gain time for their families to flee.

Cannon Hill, located southeast of Griffin's Hills but still on the north side of Battle Creek, appears to have been the location of the two Gatling guns as well as the Parrott rifle (or gun, in some terminology) in the early stages of the battle. Miles gives a sparse account of the engagement, saying that a skirmish line was deployed, cavalry battalions were employed to the left and right of the central column, then the formation swept forward. Miles reports "A general advance was ordered, the artillery opened fire, ... the troops scarcely making a halt, advancing from crest to crest, the Indians retreating take a good position only to be charged out."

Phase 1 of the two-phase Red River War project covered an area about five miles in length and two miles across. The

Parrott rifle. This light artillery piece could fire 10-pound cannon balls or canister: .65-caliber balls bound in a canvas bag. Canister is shown in the illustration. Drawing by Jason Eckhardt.

The model of Gatling gun used by the U.S. Army during the Red River War had 10 barrels of .50-cal. Drawing by Jason Eckhardt.

archaeologists believe they covered about 40 percent of the area of the battle during Phase 1. The techniques used included the use of volunteers to walk the site and the use of metal detectors from several different manufacturers. Phase 2 took place in 1999 and utilized the same methods but expanded the area for investigation. The army drove the Indians south and west, across Battle Creek and through the Red River breaks. The total area is approximately 20 miles in length and 4 miles wide.

Again, characteristic markings on cartridges and bullets allowed the movement of individual weapons to be tracked across the battle site. There were at least thirteen types of guns used in the battle, and evidence was recovered for at least seventy-one individual weapons. Several metal arrow points were recovered as well. Of the individual weapons identified, fifty-two were reported to have been associated with the U.S. Army and nineteen were associated with the Indians. The report concludes that this ratio of 2.7 to 1 in favor of the army is probably consistent with the actual ratio of firearms on the battlefield on each side. There are a few caveats as there often are in the interpretation of artifacts.

The report warns that evidence suggests the Indians relied heavily on muzzle-loading rifles that do not use shell casings. The number of muzzle-loaded rifles may be (and probably is) under-represented because the rifle balls found were not differentiated by individual weapon. Having personally fired numerous rounds from a muzzle-loader, and having examined the recovered balls, I can empathize with the difficulty of analyzing them. As the rifle barrel becomes more fouled, the characteristic rifling marks on the ball change. Add to this the variation in materials used to hand-cast each lead ball and the slight difference in size and form between hand molds, and you may readily agree that it is difficult to attribute them to individual weapons.

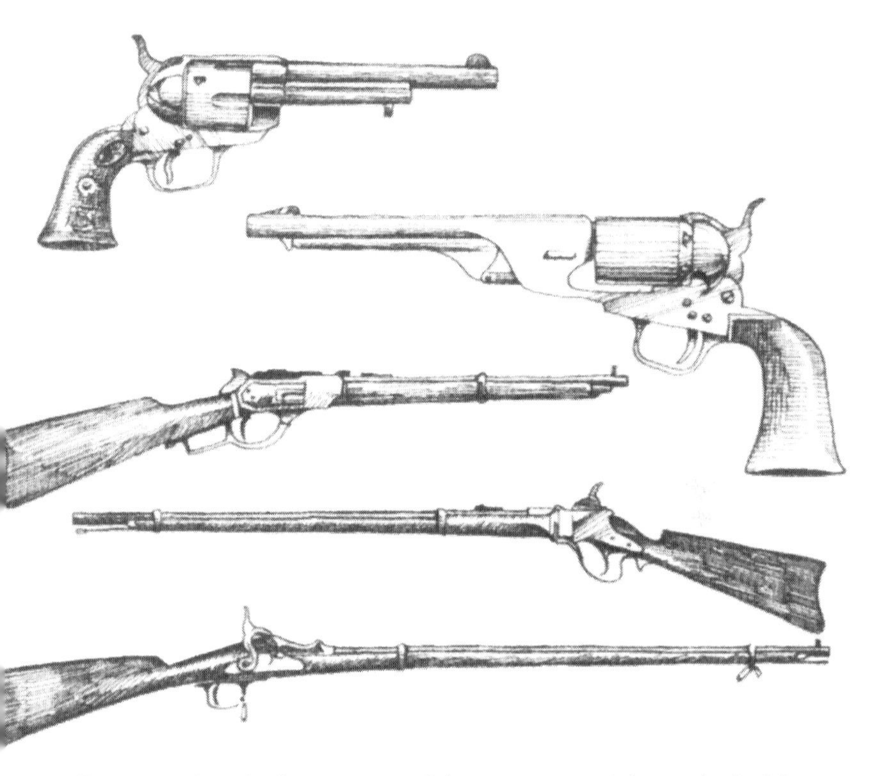

These pistols and rifles are some of the weapons used during the Red River War. From top to bottom: Colt .45 single action "Peacemaker"; Colt Army .44-cal., cap and ball pistol; Winchester .45-cal. model 1866 "Yellow Boy" (named for the color of its brass receiver); Sharps .50-cal "Buffalo Gun"; Springfield .45-70 military rifle. Drawing by Jason Eckhardt.

Indians were using older model repeating rifles, including the Henry or Winchester model 1866 lever action rifles firing Spencer cartridges. The cartridges, according to the report, were not manufactured after 1866. The conclusion is that the Indians were in the habit of recovering the shell casings and reloading them. This would certainly alter the indicated ratio. Still, in balancing the artifact patterns, the archaeologists conclude that the Indians were overwhelmed by the superior firepower of the U.S. Army and that Colonel Miles overestimated the number of Indians opposing him.

Artifacts found include:

- .44-caliber cartridge cases (Indian)
- .45-caliber cartridge cases (U.S. Army)
- .45-caliber cartridges, unfired (U.S. Army)
- .50-70-caliber cartridge cases (U.S. Army)
- .56-50 cartridge cases (Indian)
- .56-56 cartridge cases (Indian)
- .45-caliber bullets (U.S. Army)
- .50-caliber Sharps bullets (Indian)
- .54-caliber rifle ball (Indian)
- Iron arrow points (Indian)
- Rifle barrel, .50-caliber muzzle-loader (Indian)
- Rifle ramrod, Springfield (Indian)
- 50-70 Gatling gun cartridge cases (U.S. Army)
- 50-70 Gatling gun bullets (U.S. Army)
- Parrot shell shrapnel (U.S. Army)
- Awl (Indian)
- Bell (Indian)
- Buttons, military (U.S. Army)
- Bore cleaning tip (Indian)
- Buckle, miscellaneous (U.S. Army)
- Buckle, spur (U.S. Army)
- Button, other (Indian)
- Can, leaded (U.S. Army)
- Canteen top with chain (U.S. Army)
- Concha (Indian)
- Curry comb (one, prob. U.S. Army)
- Harmonica core (three, prob. U.S. Army)
- Bridle headstall plate or bridge
- Muleshoe (one, prob. U.S. Army)
- Horseshoes (prob. U.S. Army)
- Horseshoe nails, unused (prob. U.S. Army)
- Horseshoe nails, used (prob. U.S. Army)
- Jingle or coscojo (Indian)

- Rein chain (U.S. Army)
- Saddle ring (U.S. Army)
- O-ring (one, prob. U.S. Army)
- Spur, U.S. Cavalry
- Strap loop (U.S. Army)
- Trace chain (U.S. Army)
- Wagon plate (U.S. Army)
- Utensils (Indian)

Battle of Lyman's Wagon Train

The U.S. Army used its abundance of munitions and supplies liberally. One notable feature of the survey of the Battle of Red River was the discovery of a trail of 122 unfired cartridges, evidently lost from a supply wagon in its hasty attempt to keep up with the fight. The cartridges apparently dribbled out, forming a trail that demonstrates the difficulty of the chase. Broken and lost tack along with horseshoes and leaded tin cans hint at the enormous material cost of fielding such a force in 1874. Miles's column had to carry everything they needed, and resupply was difficult in the face of a hostile nomadic enemy.

Miles established his headquarters on the Red River and sent civilian scouts to Camp Supply. The scouts were conveying orders to dispatch supply wagons. In the meantime, he sent a military escort under the command of Captain Wyllys Lyman and thirty-six wagons east to meet the supply train. Captain Lyman's force consisted of sixty-six armed men. Of the sixty-six, fourteen were mounted and the remainder were infantry. The Lyman train met the Camp Supply wagons in Oklahoma at Commission Creek, transferred the supplies of ammunition and provisions from the Camp Supply wagons to their own, and set out for Miles's Red River headquarters.

On September 9 the Lyman caravan began to encounter Indians as they made their way across the divide between the Canadian and Washita Rivers. The terrain is rolling prairie composed of sandy loam soil. The Indians occupied the high ground along the trail, and Lyman eventually had no choice but to make a stand. He circled the wagons into a "corral." The position was nearly overrun by an Indian charge before it was well established. The army formed a skirmish line and was able to defend the train against the attack. Lyman ordered the digging of rifle pits on the tenth when it became clear to him that the siege would not end soon.

On the twelfth Lyman dispatched a scout to Camp Supply to summon help. He estimated they were faced by a force numbering 400 warriors, who held the high ground. He reported receiving copious, withering fire from September 9 through the 14th, and the siege continued until the Indians noted the approach of an army cavalry column, probably the 8th from Fort Union. The scout, W. F. Schmalsle, had in fact made his way to Camp Supply by the twelfth and arrived back at Lyman's position with a company of cavalry in the early hours of the fourteenth. Lyman lost two soldiers and a teamster. The casualties on the Indian side were, no doubt, much higher.

The archaeology supports the location of the battle, the positions described by Lyman in his report, and the general sequence of events. Lyman describes the corral of wagons as being "D" shaped, due to terrain features. The positions of rifle pits were quickly located by the archaeologists, and they confirm this detail. The pits were found to be rectangular with squared corners. Lyman reported the Indians retired to ridges after the initial assault. He estimated the ridges to be 400 and 900 yards from his position in a circular pattern. The Indian munitions were found on ridges at 547 and 900 yards distant. The ridges run in semicircular fashion to the north and east of the rifle pits.

Since Lyman's command spent six days under siege, the concentration of horse and mule tack, personal items of clothing, remains of leaded tin cans, glass, and cookware should be present, and they were. A total of 1,176 battle-related artifacts were recovered. The munitions were similar to those recovered at other Red River War engagements: the army was equipped with standard issue .45-55 carbines, .45-70 military rifles, and .45 revolvers. The Indians had the usual assortment of older Winchester and Spencer repeating rifles along with military and trade model muzzle-loading rifles. Bows and arrows may have been used as well, but only one iron point was recovered.

The ferocity and volume of fire reported by Lyman is not evidenced by the artifacts. The Phase 1 report quotes from Lyman's report that the Indian fire was "constant and severe" and "so sharp that we had to lie close." Lyman continues, "They must have expended a great quantity of ammunition." Accepting Lyman's estimate of the numbers of warriors, the ratio indicated by the artifacts recovered works out to nine cartridges fired by soldiers for every one fired by Indian warriors. Again, the ratio could be affected by the Indians' collecting (curating in archaeological terminology) cartridges. The entire battle site could not be surveyed, which is another factor that affects accuracy. As with other sites, there is no way to know how many and what type of artifacts were removed from the site by collectors. Even accounting for these possibilities, the pattern is still clear to the archaeologists: "If there were as many Indians at the site as Captain Lyman estimated, then, compared to the military, the Indians were not very well armed or they were being very conservative with their ammunition."

Of the 1,176 artifacts found, 892 were associated with the U.S. Army and 284 with the Indians. The military artifacts included leaded and unleaded tin cans (15), glass, cookware, and hardware such as harness pieces, and miscellaneous screws, nuts, and washers. The Indian artifacts included the usual assortment of munitions and three knives.

Battle of Buffalo Wallow

The only Red River War battle available for inspection by the general public, other than Adobe Walls, is the Buffalo Wallow site. A granite monument was erected on property deeded to the Panhandle Plains Historical Society. It stands near the junction of U.S. Highway 83 and Texas State Highway 277 on an unpaved county road. The monument was placed according to recollections of survivors of a minor clash between a group of Plains Indians (possibly Kiowa) and a detachment of soldiers and scouts who were caught in the open by a raiding party retiring from the siege of a wagon train nearby.

A survey of the area around the monument was conducted by archaeologists associated with the State Historical Commission and Panhandle Plains Historical Society. The area was swept with metal detectors and a number of rifle cartridge casings were recovered. All of them were of calibers consistent with the weapons used by the military and by the Indians of the period. The locations of the cartridges suggest the monument is placed at or near the buffalo wallow the soldiers used for cover during the attack.

On September 12, 1874, a detachment of six men was sent by Colonel Nelson Miles to locate Captain Wyllys Lyman and the thirty-six-wagon train under his command. Lyman's wagon train was to meet with another wagon train dispatched by Colonel Miles with much-needed supplies to refurbish depleted stocks. Billy Dixon, another scout, and four soldiers held off the Kiowa braves for most of the day. They made their stand in a depression in the ground, called a buffalo wallow, created by buffalo rolling in a muddy depression. The soldiers are reported to have increased the depth of the depression by digging in it with their knives. They used the earth to form an embankment around the perimeter of the wallow, giving additional protection. Two of the soldiers were killed before a rainstorm discouraged the Kiowas from continuing the assault. The one hundred twenty-five Indians could have made short work of the six men of the detachment if they had been willing to do so. Why they didn't do so is a matter for conjecture.

Billy Dixon left on foot to find help for the men, all of whom were wounded. He encountered the 8th Cavalry, led them back to the wounded men, and was later awarded the Congressional Medal of Honor for his actions.

Eight fired cartridge cases were recovered from the battle site, all believed to be associated with the Indians at the battle. Five cartridges were caliber .50-70. The cartridges gave evidence of having been fired in a Springfield model 1868 or 1870 rifle, as well as a Sharps .50-caliber rifle. Three cartridges were Spencer caliber .50-56. The markings on the cartridges indicated they were fired from three different Spencer repeating rifles. Evidence indicates the cartridges represent five different weapons.

Battle of Sweetwater Creek

On September 12, 1874, elements of the 8th Cavalry numbering 110 troopers, scouts, and wagon teamsters were traveling north toward Sweetwater Creek. Near the present-day location of Mobeetie, Texas, they encountered a party of Comanche and Kiowa warriors. The resulting four-hour running battle is known as Price's Engagement, or the Battle of Sweetwater Creek.

Major William R. Price had left Fort Union, New Mexico, on August 24 at the head of 216 troopers and additional support personnel. Price divided his command on September 4 for the purpose of establishing a supply camp near Adobe Walls. He turned south in order to locate Colonel Miles's headquarters. Price had two mountain howitzers; he sent one with the Adobe Walls detachment. Price further reduced his strength by sending twenty troopers in search of Miles's supply train. These may be the soldiers whose presence ended the siege on Captain Lyman's wagon train.

Price located Miles's camp. He conferred with the colonel then moved northeast. Between the dry fork of the Washita River and Sweetwater Creek, they spotted the Indians deployed in a line on a ridge to their southeast. Price reports that he threw out skirmishers and advanced. The troopers charged, drove the Indians from their position, and consolidated in order to repeat the maneuver. He states that while the troopers chased the Indians from ridge to ridge, always eastward, a group of warriors concentrated around his howitzer. Price says he took a platoon to the defense of the gun and successfully prevented its capture by the Indians.

Price estimated that he faced 150-175 warriors, all armed with "long range guns." He states he was outnumbered two-to-one and believed the Indians were determined to defeat his command. His account portrays a gallant and heroic battle, and that the Indians "received a very severe lesson." Price reports the fight over the howitzer but gives no indication it was used in the battle. To the contrary, he states that due to the fact that they had been drenched in a downpour, the howitzer shells were soaked and unusable.

The Indian version, given in the Battle Sites Project Phase 2 report, draws a very different picture. The Indians, warriors together with their families, were behind a ridge moving southeast. They were not visible to Price. The Indians knew the soldiers were coming, and they staged their attacks to decoy the soldiers away from their families. They mention the rainstorm, the "small cannon," and the charges made against the army troopers.

A comprehensive survey of the battle site required the cooperation of landowners along with the assistance of the Panhandle-Plains Historical Association, the White Deer Land Museum, and the Panhandle Archeological Society. The battle site was located by use of a global positioning satellite receiver and confirmed with metal detectors. The starting point was indicated by crossed sabers (with no other label) on an 1875 army map.

Once again, the sequence of events in the official reports is confirmed by archaeological evidence. The ferocity of Indian fire, once again, is not. Using Price's estimate of the numbers he faced, the ratio works out to be 3.8 to 1.5 cartridges fired, in favor of the army. The pattern of artifacts shows that the army was drawn away to the north while the Indian families escaped southwest. There is support for Price's account of the battle for the howitzer. Contrary to Price's report, there is evidence that the howitzer was used in its own defense.

Price had available both shell and canister for his howitzer. He states his shell was useless due to drenching by the rainstorm. Apparently, the canister was not so affected. Twenty-four canister fragments and eighty-three canister lead shot were found over a kilometer's distance. It is apparent from the distribution of artifacts that the Indians made a concerted effort to capture the howitzer, but their main objective was to draw the soldiers away while allowing the noncombatants the opportunity to escape. The Indians must have been hard-pressed by the tactics Price's troopers employed. This conclusion is supported by the discovery of a cluster of unfired Spencer cartridges. This particular ammunition was not manufactured after 1864 and had been conserved since that time.

770 artifacts were recovered, including:

- .44-caliber cartridge cases, fired (Indian)
- .44-40-caliber cartridge cases, fired (Indian)
- Spencer cartridge cases, unfired (Indian)
- Metal arrow points (Indian)
- .45-caliber cartridges, fired (U.S. Army)
- .50-70 cartridges, fired (U.S. Army)
- .50-70 cartridges, unfired (U.S. Army)
- Primer (U.S. Army)
- Priming wire (U.S. Army)
- Howitzer friction primer (U.S. Army)
- Howitzer canister fragments (U.S. Army)
- Howitzer canister shot (U.S. Army)
- Button (U.S. Army)

Summary

Brett Cruse, archeologist for Region 2, Archeology Division, Texas Historical Commission co-authored *Red River War, The Battle Sites Project*. He assesses the value of this archaeological project as follows: "...while the investigations have served to corroborate much that is contained in the historic military accounts of the battles, they have also served to call into question other aspects of the accounts. For example, while the military accounts repeatedly mention the large numbers of well-armed Indian combatants, the analysis of the cartridges and bullets recovered from the battle sites suggest that the number of Indian combatants may have been substantially less than the military accounts suggest. The analysis also suggests that the Indians were not as well armed as the military believed they were, as the

Indians appear to have been relying primarily on older model rifles."

Cruse's co-author, state archeologist Patricia Mercado-Allinger, was quoted by the *Amarillo News-Globe* newspaper in an Associated Press article. She asks "Could you imagine fighting against soldiers with a Gatling gun when all you have is bows and arrows?" She reported, in the same article, finding evidence of massive camps occupied by Indian women and children. The evidence consisted of toys and artifacts of domestic activities such as sewing and cooking. The camps were only a few miles from the site of the Lyman's wagon train battle site, 1874. Mercado-Allinger goes on to say, "If it was a victory for Miles, it was a hollow victory...even though they killed 17 Indians in the last battle, the entire number of women and children escaped northward. Miles and his troops never did catch up to them."

The Battle Sites Project, Phase 1 report quoted General Philip Sheridan as having said of the Red River War battles, "[They were] not only comprehensive, but...the most successful of any Indian Campaign in this country since its settlement by the whites." His contrasting opinion was no doubt framed by the fact that the Panhandle Plains was soon occupied by ranching and farming interests, that towns sprang up on or near the sites of Indian villages, and that the vast herds of buffalo were replaced by domestic stock.

Chapter Nine

The Levi Jordan Plantation Site: Study in Historical Archaeology

In 1848 Levi Jordan, along with twelve slaves, arrived in Brazoria County, Texas. They'd come to fulfill Levi Jordan's ambition to establish a plantation on the 2,222 acres of fertile coastal plain purchased by Jordan at the price of $4.00 per acre. By 1860 Jordan had built a sugar press, processed two crops of sugar cane, and become a millionaire. He kept at least 134 slaves, owned a steamship, and may have had the first sugar processing operation in Texas.

The lead archaeologist, Kenneth Brown, Ph.D., used those records available to frame the inquiry that guided the various excavations. Use of public records typifies the specialty of historical archaeology. This is only the tip of an iceberg of difference in the richness of methods and scientific techniques that distinguish the field. Living memories of descendants (of tenants and of the Jordan family) would be incorporated as hypotheses. Statements of expected outcomes were framed. Personal communications in the form of letters and diaries were used, when available, to ask more detailed and relevant questions than could be answered by excavation. Material findings and conclusions reached through excavations at other historical sites were considered. The hypotheses were balanced by null hypotheses—the results expected should the hypotheses turn out to be incorrect.

History does not guide the interpretation in anthropological archaeology, as it does in the European school. Anthropological principles, as well as all historical documents including public and private records, diaries, oral histories, and traditional stories, may be taken into account by the archaeologist in order to pose questions that can be answered through excavation. Such was the case when the owner of the site offered to Dr. Brown the opportunity to dig the Jordan plantation.

Near right: the plantation "great house." Near center: the cookhouse. Background: the slave/tenant farmers' quarters. The great house was constructed of Florida yellow cypress. Native Texas woods were used in construction of the cookhouse. The quarters were built of brick produced by the plantation slaves from local clay.

Among the anthropological questions asked and answered at this plantation were: What was the social structure employed by the occupants of the slave quarters? Were there indications of carry over from the African cultural origins? Was the society imposed by the planter, Levi Jordan, or was there an indigenous social order? Were there specialists among the quarters' occupants, and if so, what specialties are represented by the archaeological evidence?

The plantation "great house" still stands. Established as a sugar plantation, the house and lands were occupied by the Jordan family and descendants from 1854 until about 1891. The occupants of the slaves' quarters became tenant farmers after 1865 and the conclusion of the Civil War. Due to a dispute among Jordan descendants, the tenants abandoned the quarters or were forced out. The artifacts suggest the tenants left in haste: Numerous personal items, such as jewelry, coins, and pocketknives were recovered during the investigation. Since Dr. Brown has not yet published his interpretation for the site, I will forego further discussion here.

The slaves/tenants quarters were flooded soon after abandonment, and the site was buried under several inches of silt. It was disturbed only once, by a contractor who removed many of the bricks that formed the walls of the quarters. The sealed quarters remained undisturbed for nearly 100 years, protected by a layer of heavy river bottom silt from the early 1890s.

The excavation of the site under Dr. Kenneth Brown, University of Houston, began in 1986 and was completed in 2002. The plantation was the site of numerous field schools held for students of the University of Houston. The author participated in two field schools and was a unit supervisor in the second field school attended.

Historical archaeology uses feet and tenths of a foot (ex: 1.2 inches) as the standard of measure. The units were often assigned to two or more students at a time. Each unit was five feet square. Using a surveyor's transit and level, the unit was further divided into 25 one-foot-square subunits. Each subunit was excavated in turn, until the unit had been "taken down" one level, or one-tenth foot.

My occasional unit-partner and longtime friend, Carol McDavid, has given permission to reproduce an abridged version of a paper written describing the experience of excavating at the plantation. The paper that follows gives an excellent sense of what the experience was like for many of us.

Excerpts from "The Levi Jordan Plantation, an Academic Workplace, an Ethnographic Exercise" by Carol McDavid

(Note: the paper is written in present tense because the excavation was ongoing at the time, Ed.)

The Archaeological Setting

Archaeology, by definition, destroys a site as it is dug. The context is forever lost unless the site is properly dug, recorded, and analyzed. The students' work—how and how well it is done—is therefore extremely important at the Levi Jordan plantation.

The Physical Setting

The Levi Jordan plantation is located approximately 6 miles south of the town of Brazoria, Texas, on Highway 521, in the lush, tropical southeast Texas coastal area—the Gulf of Mexico is only 14 miles away, and the San Bernard and Brazos Rivers both run nearby. The summer climate is hot and damp, and the usual south Texas varmint population abounds at the site—mosquitoes, spiders, the odd snake, and most annoying of all, fire ants. During the late summer, spring, and winter, the foliage around the quarters area grows in abundance, and a big part of the beginning of every summer's work involves clearing small trees, slashing at undergrowth, and in general creating a navigable work area. The quarters lie at the rear of the property, about 100 yards from the main house, separated from the lawn around the house by a barbed-wired fence and a line of trees—many students commented that they had a feeling of "going back in time" whenever walking back to the quarters to begin the day's work.

The original plantation house is set about 200 yards back from the highway, visible from the road and marked by a Texas Historical Marker. A long unpaved "drive" connects the cattle-guard entrance from the house (the property is used by one of the owner's relatives to run cattle), and city-slicker students always have trouble remembering that the gate does, indeed, have a purpose other than security against human interlopers. Cows have been known to visit the house and digging areas, leaving unmistakable signs of their visits—the combination of cows and ants argues against the use of sandals as footwear on any part of the plantation, and bare feet are definitely not recommended.

During field schools, students' cars line the side of the property, and a large oak tree beside the house shelters a picnic table and chairs. The house is surrounded by a thick carpet of Saint Augustine grass, well populated by the aforementioned fire ants, and contains many fine old trees and a number of outbuildings.

The two-story house itself certainly does not resemble "Tara"—rather, it looks like what it is, a nineteenth-century Texas country house, with a plain front, peeling paint, and slightly sloping walls and floors. While there have been some changes in the last century—there may have once been columns in front of the house, and a room in the back was added—in general the house is very much like it was when occupied by Levi Jordan and his family. The owner allows students to live in the house during the summer field school, but the fact that the house is neither heated nor air-conditioned and has a sanitary system that confounds those who have never dealt with the peculiarities of a septic tank, provides a self-limiting effect on the number of students who decide to "sleep over." There is, however, electricity, a refrigerator, and plenty of room, so as the summer wears on, more and more students decide to join "the house dwellers" (the relationship between house dwellers and commuting students will be discussed later) for occasional overnight visits. When the breeze is right, the grass and the trees keep the temperature at the site reasonably bearable, even in the heat of the summer. At night, the hum of electric fans competes with the chirping of the crickets, and students sit under the big oak tree until late at night, talking, eating, and resting from the day's work.

The work itself takes place primarily in the quarters area, and excavation so far has revealed that there were originally four blocks of slave quarters, each consisting of two 80' x 20' buildings. Each set of buildings was connected by a breezeway of some sort, and was subdivided into four cabins per building, for a total of eight cabins per block. It appears that each block probably contained eight cabins, although there is some indication that at least one block was subdivided into larger cabins, possibly creating a "dormitory" for young unmarried adult males. At the rear of each building, there is a large, round depression. These depressions are thought to be the remains of open cisterns used by the slaves and tenant farmers. Most of the excavation has taken place within the areas identified as cabins, with the exception of some work done last summer on what may have been a butchering or trash area at the rear of the quarters area, and work done during a previous season on an abandoned cistern next to the main house.

The Work

To take advantage of cooler early morning temperatures, the summertime workday begins at 7:00 a.m. and lasts until 2:00 p.m., with a 30-minute break for lunch at noon. At other times of the year, the start times vary: This spring we did not begin until 8:30 and stopped at about 3:30. Because the site is a long drive from Houston (an hour or more), there is generally one day per week in which driving to the site is optional—students can, if they prefer, work in the university archaeology lab processing artifacts, rather than driving to Brazoria to dig. This study will focus on work at the site, not in the lab.

Students are expected to provide and keep track of their own digging trowels, line levels, journals, and personal items such as mosquito repellent and sun screen, while the anthropology department provides the myriad of other tools and supplies necessary. A sampling of tools required includes a transit for measuring elevations, plastic and paper bags for artifacts, markers to identify the bags, plastic sheeting to cover units between excavation periods, rulers, string, files for sharpening trowels, forms and graph paper for detailed notes and drawings, assorted brushes, root cutters, first-aid kits, buckets, screens, and in the summer, gallons and gallons of ice water to prevent dehydration. Every day each student is expected to help carry the day's tools and supplies from a storage area in the main house to the digging area, and to help load everything back in at the end of the day.

The summertime dig in 1991 took place in four different areas of the site (Blocks One, Three, Four, and the previously mentioned butchering area) with a graduate-student supervisor in each who was responsible for his or her "crew" of workers. There were 20+ students enrolled in the summer group, in addition to several graduate students. Dr. Brown circulated from area to area, supervising the graduate students' work, providing help and decisions, if needed, as well as participating in the digging itself on most days. Most areas developed nicknames; for instance, the area in which I worked was known as the "corral" because a portion of it had been used as a corral sometime earlier in this century, while Block Four, with no tree shelter, was known as the "electric beach." Students in each area came to identify with their own area, and, unless a problem arose, each student usually worked in the same area for the entire six-week period. The dig this spring was much smaller—excavation only took place in one area, with a group ranging from three to six people at any given time.

The author surveys the "corral" in 1993. Photograph by Ruth Marcom, 1993.

In order to explain the students' role in the digging process, it is necessary to first briefly describe the way the site is organized. The site is divided into four quadrants (NW, NE, SW, and SE) with a zero point in the center of these quadrants. The site is further divided into a grid of 5' x 5' squares: these squares, known as units, form the basic unit of excavation. The grid coordinates of the northwest corner of each unit become the numerical designator for that unit. For example, to say that a given artifact was found in Unit 5S/55W means that this 5' x 5' unit is southwest of the center point of the grid. The northwest corner of this unit is the precise point on the grid measured as 5S/55W. At times, the 5' x 5' units are subdivided into smaller 1' x 1' squares, and when that occurs, each 1' x 1' is assigned the appropriate NSEW coordinates.

In addition to the NSEW provenience designates, elevation measurements (in tenths of feet) provide a point of reference for the precise depth of any unit dug. Before a unit is opened, the elevation of the highest corner is measured, using a surveyor's transit. This measurement becomes the starting point for continuing depth measurements as the unit is excavated, as the hole gets deeper and deeper. These measurements are made by the student, using a string, a line level, and a hard wooden ruler. Generally, levels are dug in 1/10' increments, with each level being completed across the entire unit, or subunit, before further digging is done. A high degree of precision and accuracy is required in order to ensure accurate provenience control and to enable soil changes and other features to be seen clearly.

Therefore, the students learn to keep up with two different kinds of measurement — *across* the unit (a spatial measurement) and *into* the unit (a temporal measurement; as one goes down, one goes back in time). Students are expected to be able to figure out the NSEW coordinates for any unit or subunit they dig, and to label their finds

This photo shows two units: the unit to the right exposed only brick rubble; the next unit, to the left, exposed the wall trench and the floor of a cabin. These units were in the area known as "the pasture." Photograph by Robert Marcom, 1993.

accurately and completely. They are also required to dig carefully and precisely, stopping each level at exactly the right place, according to their line-level measurement.

The issue of levels becomes a very important one during the course of the excavation. When a student is first learning, the supervisors are fairly gentle with their reminders "not to go too deep"; the technique is not an easy one to master, and it takes about two weeks before most students are able to gauge their depth properly. There is a great deal of good-natured teasing of students who are having difficulty. In the "corral," one student in particular had a difficult time learning how to do it, and he is still referred to

Extensive excavation of the slave/tenant farmer quarters at the Levi Jordan plantation. The bricks seen in the foreground were hand-made by the plantation slaves. Photograph by Robert Marcom, 1994.

as the "trench digger" by other "corral" workers. Because of the importance of context, as described earlier, these measurements are crucial, and I and other students felt considerable pressure to make them properly.

The role of graduate students in the teaching process is a critical one. In addition to learning about levels and coordinates, other techniques must be learned as well: how to prepare a new trowel, how to hold a trowel, how to sit so as not to damage the surface of the unit, how to dig neat corners and walls, how to label bags, how to watch for soil changes and other unusual features, how to dig past the brick rubble while still maintaining the right level, and how

A wall footing with the bricks still *in situ*. Photograph by Robert Marcom, 1993.

to dig as quickly as possible without sacrificing accuracy. Generally, after an introduction to the site and an overview of the history and objectives of the excavation, Dr. Brown left the one-on-one teaching of specific techniques to the graduate students in each area, as he regards site supervision as an important part of a graduate student's training (Brown, personal communication, 1991). He would correct or expand upon his supervisors' instructions occasionally, but it appeared to me that such correction was not usually necessary. The students had many questions about the site—the Jordan family history, how digs went in the past, and the like—that he did not cover in his introduction, and I found that this sort of "lore" was most easily obtained from other graduate students.

Different crews worked in different areas, each area consisting of a varying number of 5' x 5' units. Each unit was dug by a team of workers assigned by the supervisor at the beginning of the season; teams tended to continue working together if they were happy and productive; on a few occasions either the graduate supervisor or Dr. Brown would switch workers to other teams if it appeared that productivity would increase by doing so, or if a worker requested a change.... After a couple of weeks, workers tended to identify with "their" areas of the site. Jealousy sometimes arose when one area or another was finding more "neat stuff," even though, as would-be professionals, many of us tried to avoid those kinds of competitions.

During the springtime dig, ...as a teacher/supervisor, I was responsible for the dig in Dr. Brown's absence and had to make decisions about how and where to dig, in addition to teaching three new undergraduates about the digging process—measuring, leveling, and so on. I also was responsible for being the official note-taker. Detailed notes are taken for each 1' x 1', 1/10" deep square that is dug, and, because they are the official record of things taken from the ground as they are discovered, these notes are a critical part of the interpretation process that takes place later in the lab.

There was no competition for "neat artifacts"—we were primarily concerned with identifying architecture in preparation for more extensive digging to be done this summer. Finding a nail was a big event, and remembering what it was like to dig for hours with no result, I tried to encourage the other students to be excited about the whole process, not just finding things.

...Experienced workers have told me that every field school is different. Not only does the cast of characters vary, but the weather, the archaeological results, and the tempo of the department at the time all have an effect on the kind of season that occurs. I suspect, however, that in-groups and out-groups will continue to develop, that some students will still make noise late at night and irritate everyone else, that new rituals and arguments will supplant the old, and that someone will still forget to clean out the refrigerator.

Summary

On a personal note, the experience of exposing, extracting, and handling the personal effects of these tenant farmers, ex-slaves, and their families is difficult to convey. I would like to say it was an academic and interesting experience, which it most certainly was. The relating of the depth of the impact of my experience requires some forbearance on the part of the reader.

Sometimes the actions of real people were revealed in the position and association of the artifacts. Tobacco pipes, religious articles, with both African and Christian contexts, sewing items, and valuables gave hints of the daily life and the daily practices of the people who made their home at the Levi Jordan plantation. The majority of artifacts were, no doubt, refuse. This is characteristic of nearly all archaeological excavations. But presence of items obviously personally important, or universally valuable, gives a definition and texture to the lifeways of these people often missing from archaeological investigations. Who would go away from a home and leave behind pocketknives, store-bought jewelry, coins, or hand-carved shell cameos? It is no wonder then that the Levi Jordan site has been referred to as "the Pompeii of America."

Artifacts found include (historical archaeology, K. Brown, D. Cooper, 1990):

- Bone fragments
- Ceramic fragments
- Porcelain

- Glass fragments
- Cutlery
- Tobacco pipes
- Munitions
- Melted lead
- Beads
- Store-bought jewelry
- Buttons
- Sewing equipment
- Coins
- Shell artifacts
- Carved bone
- Pocketknives

Chapter Ten

The American Indians in Texas

No population on earth has more written about it, nor is any population more misunderstood, than the American Indian. The different Indian societies that spanned geographical regions and were of different cultural groups were similar in the way they used the environment and in the technologies they invented. Modern Anglo-American society has translated the similarities, through fascination hampered by ignorance, into the stereotypes seen in the entertainment media.

The pendulum swings throughout history. Sometimes romanticized as the Noble Red Man, sometimes vilified as a hopeless savage, always regarded as an impediment to the divine right of white dominion, the accomplishments, achievements, and knowledge of the Indians has usually been misinterpreted.

The complicated truths, disguised by the myths and tales, are the province of history, anthropology, and of anthropological archaeology. Archaeological investigations and anthropological ethnographies have revealed a rich tapestry of cultures, societies, and religions. There is little, if any, doubt that the societies found by Europeans upon arrival were equivalent to European culture in many respects.

The Mississippian culture of the eastern U.S. rivaled the Aztecs and the Egyptians in their civilization. They are represented by the Caddo in East Texas. The Puebloan culture of the Southwest has Antelope Creek people in the Panhandle and other tribes in far West Texas. The Plains Indian tribes, the desert tribes, the mountain tribes: all were as foreign to each other and as different in their lifeways as they were different from the Anglo/Hispanic/Franco Europeans.

Many cultural traditions came together in Texas. Coastal environments, hardwood forests, southwestern and chaparral deserts, mountains, high plains, coastal plains...every environment dictated a different strategy for making a living. Every environment attracted and shaped a people to inhabit and extract a living from it. Every single one of those peoples created and refined a culture that supported a society, generation after generation.

American Indian cultures were supreme and unrestrained on the North and South American continents for at least 14,000 years, minus the 460 years since contact with Europeans. Their skillful exploitation of the land and their use of the plants and the animals were envied and emulated by the newly arrived colonists, even while those colonists were doing their best to eliminate the Indians. The resentment and anger felt by sovereign peoples, invaded, killed by disease and violence, subjugated and dispossessed, can be easily understood.

A few sites representative of Indian Nations have been discussed previously. Taken individually, they fail to represent the interlocking, sometimes adversarial, sometimes cooperative, always timeless and universal sense of Indian culture in North America, and in Texas. A larger perspective would include a larger scale and a longer perspective.

Three *culture areas* intersect in Texas: Southwest, Plains, and Southeast. These culture areas are the result of work done by early anthropologists and are defined by the ethnic similarities of neighboring groups. Archaeologists use another classification, *tradition*, to indicate a set of artifacts, including tools, building styles, art styles, and pottery, to indicate a persistence of culture through time. For instance, the Panhandle and West Texas were influenced by the Anasazi and Mogollon traditions of the Southwest. East Texas has been dominated by the Mississippian tradition. Some evidence is put forward for inclusion of several coastal groups, including the Karankawa, in the Caribbean tradition. The Desert tradition is well established by the Coahuiltecans in South Texas and northern Mexico.

The science of linguistics offers another method for classifying Indian cultural continuity. Language is arguably the most basic and the most important human cultural tool. Language groups are evidence for the history and descent of different ethnic groups. As might be expected, a variety of language groups are found in Texas: Athabaskan, Hocaltecan, Azteco-Tanoan, Uto-Aztecan, and Caddoan are the main groups, with Muscogean, Souan, and Algonquian added in historical times.

These classifications are useful because the names of Indian nations, tribes, and bands change over time and are recorded in a number of Western European languages. The Spanish and French made first contact with Texas Indians and often gave different names to the same groups. Additional confusion crept in when the same group was given different names as they were encountered in different areas. Attempts have been made to reconcile the conflicting names, but the identity of many of the groups recorded by early European explorers will remain mysterious. The tribes of Texas, from late prehistoric times to present, are only tentatively identified. Prehistoric, Archaic, and Paleoindian cultures and traditions can only be known by their artifacts.

Nations, Tribes, and Bands

The intersection of historical documents, anthropological ethnography, and archaeological interpretation has given us a window on the people of Texas, both before European contact and since. The names of Indian nations, tribes, and bands used here are those I found to be most common. Since this book is not intended to be a scholarly treatment, I will distill and interpret the information available to me. Casual research will reveal that I have sidestepped a number of controversies and conflicting opinions in the interest of concise description. If the field of paleoethnology is interesting to you, I invite you to plunge into the vast ocean that we will skim in the remainder of this chapter. For a more comprehensive treatment of many of the tribes and nations, designed for the lay reader and for students as well, I recommend *The Indians of Texas*, by W. W. Newcomb Jr., University of Texas Press.

Southeast Culture Area — East Texas, from the Red River to the coastal plain

The countryside varies from dense pine and hardwood forest to open parkland. East Texas is well watered by several rivers and many streams. Vegetation is profuse, and at its southern reaches includes cypress swamps, patches of jungle, and bayous.

Caddo — Hasinai, Kadohadacho

When the Spanish and French entered East Texas, they discovered about two dozen tribes associated together in an alliance known as the Caddo Confederacy. A northern group of tribes centered around the area south, and after its southern bend, east of the Red River. Another group of tribes occupied the area south of Tyler. They possessed their own language group, Caddoan, which included minor variations of dialect for the main language, as well as Pawnee and Wichita variants, which were mutually unintelligible. Texas derives its name from the Caddoan word for friends, or allies, "teshas." The Spanish spelling of the word utilizes the "x" for the "sh" sound: texas.

Caddos grew tobacco, beans, squash, sunflowers for seeds, and the staple crop, maize. They had two varieties of corn, including a summer-fruiting hybrid and a fall-fruiting variety. They ranged far and wide, especially after acquiring the horse, and frequently engaged in buffalo hunting, using dogs trained to drive the buffalo. They hunted a wide variety of small game and harvested wild berries and fruit.

Caddo practiced matrilineal kinship, and some women had great authority. Caddoan men participated in many types of work that would have been spurned by men of other tribes. Marriage was of a casual monogamous form. Men occupied the bureaucratic offices of leadership.

Caddos practiced a "hit and run" form of warfare, similar to other Texas Indian tribes. They took scalps, raided for horses and women, and practiced a form of ritual cannibalism similar to that practiced by their relatives, the Wichita and by unrelated tribes like the Karankawa and possibly the Bidai. A captive would be displayed for days, tortured, ritually butchered, and parts eaten with ceremony.

Caddos were multitheists but worshiped a primary deity, *Caddo-Ayo*. The top leadership consisted of a secular chief, called *xinesi* and a priest. Their religious activities focused around temples built on a mound. The temples were constructed in the same manner as their houses, pole and thatch, but were larger. They practiced shamanistic medicine and used plants and soils to treat illness. They supported occupational specialists and had a number of societies, including a homosexual society of *berdache*.

Atakapa — Bidai

The Atakapa occupied the coastal plain, from Louisiana to East Texas. They had a distinct language, a form of agriculture centering on maize, and harvested clams and fish from canoes. The Atakapa are poorly known, being largely overlooked by the French and Spanish explorers and possessing little of interest to Anglo-European settlers.

It is possible that the people who gave refuge to Cabeza de Vaca, whom he called the Hans, were Atakapan. Several European explorers gave account of the Bidai, noting they had "large heads" (possibly shaped by being strapped to headboards in infancy) and dark skins. They were represented as unattractive, smelly, and cannibalistic. Their houses were comprised of sticks and brush, their canoes described as fragile, and their character as honest and peaceful.

A fair description of the Bidai might be more generous, taking into account the fact that they lived in an environment often damp and humid and filled with voracious mosquitoes. They were described as good hunters and growers of a "superfine" maize. They quickly formed alliances with the Spanish. After they were decimated by disease, they joined other, more viable tribes including the Caddos and the Alabama-Coushatta.

Cherokee

Cherokees arrived in Texas around 1800. One of the "Five Civilized Tribes," the Cherokees built towns with viable businesses, adopted white culture, and became prosperous and politically integrated by the early 1800s. In 1830 President Andrew Jackson, a veteran Indian fighter, forced the Cherokee, Choctaw, Creek, and Chickasaw tribes from their homes in the east—ignoring a U.S. Supreme Court decision against the forced concentration on reserves. Some did not go quietly on the trail of tears. A group of refugees, largely Cherokee but including individuals from the other tribes, petitioned the Caddo for a home in the Confederacies. At first they seemed to have a possibility to avoid the prejudice that cost them their homes in the east. They made alliances with Sam Houston and the Republic of Texas. Unfortunately for them, a politician of Jackson's ilk, Mirabeau Buonaparte Lamar, became the president of the republic. The Cherokees were forced to leave Texas in 1836, relocating to the Indian Territories.

Alabama-Coushatta

Between 1780 and 1800, members of the Creek Confederacy, of the tribes of the Alabama and Coushatta, began to arrive in East Texas. They had a cultural affinity to the Caddos, and being known to them, the Caddos gave these refugees a home. The Alabamas and Coushattas were first said to be living on the Alabama River in 1541 by Hernando De Soto. They found themselves under French authority after France laid claim to the Louisiana Territory. When the English intruded into the area in 1714, the Alabama and Coushatta began migrating east, following the dominion of the French.

They arrived in Texas circa 1780s and settled in the vicinity of the Sabine River. They found the Spanish eager to recruit them for patrols. They were charged with the duty of keeping French trade missions out of East Texas. The Spanish were lavish in their payment for services, and the country offered an abundance of game. They became the rulers of the Big Thicket and prospered.

Today the Alabama-Coushatta are one tribe, occupying 4,600 acres in the state of Texas. Because of their history of cooperation, and through the efforts of Sam Houston, they were granted a reserve by the Republic of Texas of more than 1,800 acres. They have the oldest, largest, and only remaining reservation in Texas.

Plains Culture Area — Includes coastal plains, southern plains, and high plains

This sea of tall grass extends from southern Texas into southern Canada. Home to enormous herds of bison, deer, antelope and a variety of smaller game, the "Great Plains" have been a mecca for hunters for 14,000 years. Special skills were required for making a living from hunting on the prairies. Water and game went hand in hand, and often, especially in times of drought, skilled scouting was necessary to avoid starvation.

Powerful occasional storms roar across this expanse of emptiness, both winter and summer. Lightning strikes set fires, in earlier times, that incinerated tens of thousands of acres, running from horizon to horizon. Until the horse came to the Plains Indians, life on the plains could be a very risky proposition.

Tonkawa

The Texas plains, while generally flat and grassy, offer several different environments. The southern plains are home to the Tonkawas. These people are thought to have arisen from a local archaic culture, possibly the Toyah focus. They were known to the Spanish and the French through some sensational tales of cannibalism written by explorers and missionaries. The Anglo-European settlers of Texas knew them in general terms as benign and peaceful and not particularly noteworthy. Accounts by Texans are often dismissive or even contemptuous of the "Tonks."

The southern plains of central Texas was the winter home of the great southern bison herd. The Tonkawa aspired to make the main focus of their subsistence the bison hunt, and in prehistoric times, they most likely succeeded. There are stories told by explorers of seeing bands of Tonkawa with travois, made of two poles between which skins are stretched, being pulled by dogs. The Tonkawa used the bow and arrow and sometimes a short spear as weaponry. Their tipis were squat; sticks were leaned together and secured at the top. They were draped with hides and were regarded as crude, compared with the grand, well-crafted tipis of the northern Indian tribes.

It is not possible to verify the tales of Tonkawa dining on human beings for subsistence. They certainly did practice ritual cannibalism, as did most of the tribes in Texas (excluding the Jumano and the Comanche). Their main source of meat was probably small game. They gathered nuts, fruits, berries, and unlike other Plains Indians, harvested fish and shellfish.

Tonkawas adorned themselves with shells, horns, teeth, and all manner of items. The women were noted to tattoo and paint their bodies with elaborate geometric decorations. The most common design was one mentioned repeatedly: concentric circles on each breast, beginning at each nipple. The men wore leggings, moccasins, and whatever other article of clothing that was dictated by the weather. Women wore a short animal skin skirt and complemented it according to the demands of the climate.

The Tonkawa were frequent visitors to missions and later, to forts. They came to trade, offering hides and meat from wild game. Being less numerous than their primary rivals, the Comanches, they became "mission Indians" whenever it was convenient or necessary for safety. They were often called lazy by the Europeans, being reluctant to become involved in agriculture, mining, or other industry. The truth is, they lived in a well-watered and fruitful part of Texas, beginning on the Edwards Plateau, extending south into the Brazos River Valley. Hunting and gathering in such an environment required little effort, and the life insisted on by the Europeans must have looked onerous and pale by comparison.

The Tonkawas were purged from Texas during Lamar's campaign to eliminate all Indians from the republic.

Comanche

The Comanche Nation was born on horseback. The origins of this most feared and respected of all Indian tribes is only obvious when the Comanche language is traced to its roots. It is virtually identical to the language of a northern mountain tribe, the Shoshone (Uto-Aztecan). On further inspection, much of the Comanche culture and society can be found rooted in the Shoshone, as well. The Comanches became a distinct culture after the Western (Desert) Shoshone became mounted.

Sometime in the early 1700s, the Proto-Comanche bands began to hunt and live on the northern plains. They ranged further and further south, until they found and began to raid Spanish and French missions in the mid-1700s. Their reputation among other Native Americans was that of being skilled thieves (an honorable trait), brave warriors, and implacable enemies. Often called "the finest light cavalry ever to sit a pony," they drove out the Apaches and the Tonkawa and forced alliance upon the Cheyenne, Arapaho, and Kiowa.

By the time of the days of the republic, they'd become public menace number one for the new nation. They were not a militaristic people, however. Their reputation for savagery and skill at massive raiding was largely a construction of the Anglo press urged on by a government needing to promote popular support against the Indians. The Comanche had no military societies or any large organization. Each band was an autonomous unit, and warriors joined war parties according to their own appraisal of whether the booty would be worth the effort.

The Comanches were described by Europeans as tall and attractive from their earliest encounters. Originally Comanche men and women dressed plainly, modestly covered with animal-skin clothes. Men wore headdresses featuring bison horns. Later they adopted clothing similar to other Plains Indians: Leather leggings, breast plates, beaded and quilled leather shirts and dresses were common. After contact with whites, it was not uncommon to see Comanches sporting western garb and hats—even to the point of warriors wearing women's dresses, stolen from white victims.

Young Comanche women often took part in hunting parties, successfully competing and making kills. Labor was divided according to sex, but much latitude was given to the young. Comanches were usually tolerant, indulgent parents. According to several sources, Comanches did not punish children physically, preferring to use moralistic stories and persuasive arguments. As children grew into preadolescence, the boys and the girls were encouraged to segregate—the boys forming loose bands of association and the girls attending to the women's work at home.

The nomadic lifestyle of the band was centered on following the bison herds. Comanche men carried bow and arrows, shields, long spears, a war axe, and knives. The spear was used both for hunting and for making war. The band moved on horseback, in-line with the men in front. They carried their tipi poles and skins with them, and the women traveled mounted, astride. When they camped, tipis would be positioned with the entry facing away from the prevailing wind. Women cooked, tended to babies, and gathered firewood. Men tended the horses (sometimes possessing large strings) and weapons. Meat was often eaten raw, but the women knew how to preserve it by jerking or by mixing it with pecan meal to make pemican.

Comanches lived in a world of supernatural wonders. Supernatural powers were obtained to help cope with forces otherwise beyond their influence. Power could be obtained through visions induced on a quest, from animal spirits, or by a shaman ritual, such as the Eagle Dance. Men often avoided power, because it carried great obligation with it, to serve the highest interests of the band. Women could achieve power, either by association with a husband who had power or through magic. Power was an individual attribute but could be shared upon request. The Comanche universe was presided over by a Great Spirit. Comanche spiritual life was highly individualistic. The sun, moon, and earth were obviously mystical and often figured into beliefs. They had no corresponding Spirit of Evil, and they did not fear the spirits of the dead.

The Comanches were signatories to the 1867 Treaty of Council Camp, along with the Kiowa and the Apaches. The treaty guaranteed them, among other things, the right to hunt buffalo on the southern plains, providing they move to a reservation in the Indian Territories. They were offered annuities and guarantee of provisions. When the provisions did not materialize, they returned to hunting buffalo, as was their right under the terms of the treaty. The Comanches did not consider, at the time of the signing of the treaty, that a systematic effort would be made to exterminate all the buffalo. That is precisely what the U.S. Army and General Phil Sheridan had in mind. Some of the Comanche warriors decided to attack the whites who encroached on their hunting grounds at Adobe Walls. The army responded in force—not to remove the white interlopers but to destroy the Comanches.

The Comanche bands were defeated by the U.S. Army in 1874 in a number of engagements termed the Red River War. They were forced to accept exile to the Indian Territories for a second time. Their half-white war chief, Quanah

Parker, surrendered the Kwahadi band at Fort Sill on June 2, 1875. He came to the end of his life with some degree of celebrity, being recognized as a tough-minded businessman.

Kiowa

Origins of the Kiowa are unknown. Their language is rooted in the Tanoan group, usually associated with the Pueblo Indians, the Tewa, and the Tigua, of the Rio Grande Valley. It is possible the Kiowa wandered north in prehistoric times, then returned south. By the early 1800s they were completely adapted to the high plains. They were driven out of the Dakotas by the Lakota Sioux in the 1780s; they made an alliance in 1790 with the Cheyenne and later with the Comanches after a year of intense fighting. The Kiowas and their near relatives, the Kiowa Apaches, made their living as traders of Spanish horses among the Plains Indian tribes. They were known as fierce warriors and staunch allies by the Comanche, Southern Cheyenne, and Arapaho tribes.

Two Kiowa chiefs, Satanta and Big Tree, were tried in 1871 for murder. The Warren wagon train raid resulted in the murder of the wagon master and six drivers. Both chiefs were sentenced to death. The sentences were later commuted, but the chiefs were rearrested for parole violations after they participated in the Second Battle of Adobe Walls in 1874. This raid, the parole, and the resultant rearrest of the chiefs is said to be the reason for General Sherman's change in policy from defense against Indian raids, to an active campaign against all Indians.

The Kiowas were forced onto reservations after the Red River War. They ceased to be a tribe when they, of their own volition, acculturated and integrated into local communities in Oklahoma. Many descendants of the Kiowa now live in the area of Anadarko, Oklahoma.

Arapaho

The Southern Arapaho Indians made their home in Colorado, ranging into Texas on occasion. They followed a Plains Indian lifestyle that included buffalo hunting, raiding, and horse stealing/trading. They participated in the Second Battle of Adobe Walls. The Arapaho were probably neighbors of the Blackfoot Sioux in prehistoric days.

Wichita

The Wichita people are related to the Caddoan and lived on the Blackland Prairies of northeast Texas, Oklahoma, and Kansas. They speak a dialect of Caddoan (one of four main dialects: Caddo, Pawnee, Kichai, and Wichita) and probably originated in the vicinity of the southern Red River. The Wichita were probably a nomadic people until they acquired the horse, which allowed them a greater range for hunting.

The four subtribes of the Wichita are the Wichita proper, the Waco, the Tawakoni, and the Kichai. They are described as short, stocky, and dark-skinned—as are the Caddo. They were often heavily tattooed and made extensive use of body painting. Like the Pawnee, the Wichita were quite willing to serve Anglo-European interests for pay.

They practiced a form of agriculture since their pre-horse days, growing maize, beans, pumpkins, and gourds in gardens. They were dry-land gardeners, depending entirely on the rains. According to early reports, they did not eat fish or shellfish and supplemented their garden produce by hunting.

They reportedly lived in houses made of forked sticks, bound together and covered with grass. Some of these houses were quite substantial and were used for a number of years. In later years the Wichita were forced to move into larger villages. They began to raise cattle and farm intensively. As population pressures increased, they began to enslave captives instead of torturing them. They practiced an informal ritual cannibalism on captives prior to European contact, but eventually replaced the custom with forced slavery.

Their religion was a pervasive animism, founded on the belief that everything possessed a spirit. They thought they could influence the spirits both through direct appeal (powers) and through the use of spells (magic). Early Europeans attributed the Wichita bravery in battle to the warriors' belief that they would continue to live after death.

Coahuiltecan Tribes

Aranama

These Indians were well known to the early Spanish and French explorers of the area between the San Antonio and Guadalupe Rivers. The Aranama tribe was composed of hunter-gatherer bands of the Coahuiltecan language group. They were poor in material wealth and opportunistic in their relationship with others. They were capable of turning to violence if they thought it safe to do so. Their lithics industry consisted primarily of choppers, scrapers, and a triangular projectile point. Although the point is found in a range of sizes, the size may have to do more with the chert sizes available. Some of the bands reportedly made use of the bow and arrow and even hunted the occasional bison that strayed too far south. The life of even these, the more prosperous of the Coahuiltecan-speaking Indians of South Texas, was exceedingly harsh.

The men, and probably the women as well, were tattooed and wore little clothing, other than that necessary for protection from the elements.

Karankawa

The coastal plains and barrier islands were home to these hunter-gatherers. They used spears and dugout canoes to hunt fish. They harvested shellfish and hunted every type of edible animal, as well. The Karankawas, or possibly their cousins, the Copanos, were the Indians who found Cabeza de Vaca and his crew and provided them with food and shelter (if capture and enslavement can be regarded in such light). The Karankawas were the Indians who attacked Fort St. Louis and carried away three children as captives. When the Spanish encountered the band and recaptured the children, they'd already been tattooed in Karankawa style. Karankawas made liberal use of tattoos and paint. The women reportedly wore modest dresses of deerskin and Spanish moss. The Karankawa men dressed and painted themselves elaborately for warfare.

Early explorers described the Karankawa as being universally bad tempered, harsh, and capable of enduring great ordeals. The Karankawa were expert with the longbow, their customary weapon.

Southwest Culture Area — Portions of South, West, and Panhandle Texas

Jumano

Two distinct groups of South Texas Indians were named the Jumanos by Spanish explorers—as well as several tribes in New Mexico and Arizona—but there is some evidence that the two Texas groups may have come from a common ancestry. One group lived in the Rio Grande valley and practiced an agricultural lifestyle. The other group were seminomadic hunters in the vicinity of the Davis Mountains. Little is known of the nomadic Jumanos, but excavations have revealed that the agricultural tribes were practicing

Puebloan-style gardening and lived in typically Puebloan houses made of adobe.

The Jumanos apparently settled in the Rio Grande Valley some time after A.D. 1200 and were still there during the first hundred years of Spanish exploration and settlement. They had been largely absorbed into the Mexican culture by the time of the founding of the Texas republic.

Apaches

Rivaled only by the Comanches, the Native Americans called the Apaches conjure the most fearsome images of terror on the Texas frontier. Popular literature, TV, and movies have depicted the Apaches as implacable warriors, constant enemies, both cruel and merciless. The true relationship between European invaders and the various Apache tribes and bands was rather less dramatic and much more complex than the popular mythology indicates. That the Apache Indians of Texas attacked Anglo settlers is beyond dispute, but the fact that Apaches fought alongside both Spaniards and Texans is too-often ignored.

The historical-era Apache tribes probably descended from the Indians identified as Querechos, first noted by the Spanish explorer Francisco Vasquez de Coronado in A.D. 1541.

Lipan Apache

In prehistoric times, the Lipan Apache (probably the same people as the Querecho) lived a semisedentary lifestyle in villages, which the Spanish termed *rancherias*, farming unirrigated garden plots. They left the *rancherias* during the fall to hunt bison, living in small camps, returning in the spring in time for planting.

When the Comanches came into their area of South and West Texas during the eighteenth century, the Apaches found themselves unable to sustain their *rancheria*-based lifestyle. Some moved into the mountains of New Mexico to pursue a Puebloan lifestyle. Some joined the Kiowas to become Kiowa-Apaches. The Lipan adopted a plains-style nomadic existence as Plains Indians. The Lipan Apaches subsisted by hunting—bison when possible and smaller game as opportunity presented—and by harvesting nuts, fruits, roots, and insects.

They came into direct conflict with Anglo ranchers when, as bison disappeared from the plains, they began to take cattle, both wild and domestic, in their stead. The Lipan continued to garden whenever circumstances allowed, and they enjoyed a brief respite from the Comanches through an informal alliance with the Spanish and later with the Texans, against their mutual enemy. The idea of "burying the hatchet" comes from a ceremony practiced by the Apaches during a peace conference with the Spanish at San Antonio in 1749. War hatchets and other weapons were buried as a sign of peace.

Warrior values were strongly emphasized among the Lipan, but warfare was a loosely organized, voluntary affair. Lipan warriors often joined with and participated in Texas Ranger campaigns against Mexican and Comanche foes. Flacco the Elder and Flacco the Younger were father and son Lipan chiefs who allied with Texans in numerous campaigns. Flacco the Younger fought alongside Captain Jack Coffee Hayes from 1840 to 1841. The death of Flacco the Younger in 1842, possibly at the hands of "whites," is often cited as the reason for deterioration of relations between the Lipan and the Texans. By 1874 Lipan Apaches had been driven from Texas.

Mescalero Apache

These bands were the poor cousins of the Lipan Apache, although they have fared better than the Lipan in the long run, receiving a large federal reservation in southeast New Mexico, in the Sacramento Mountains. Most of the activity of the Mescaleros in Texas was limited to raiding parties. Pressure from the Comanches and other tribes forced the Mescalero out of their original home in far West Texas in the eighteenth century. Prior to moving onto their federal reserve, they seem to have lived a hunter/gatherer lifestyle similar to the Indians of the Archaic period. Their reservation was granted by U.S. presidential order in 1873.

Summary

Most of, if not all of the Indians who came to make their home in Texas were immigrants. In this regard they are much like other, later citizens. The suggestion that Indians were more ecologically responsible than European immigrants, or that they were more noble, is equally as unfounded as the notions that they were incompetent to govern themselves or that they were savage enemies that could only be defended against through a policy of genocide. The worst fears of the white invaders seemed borne out by the all-too-frequent attacks, rapes, and murders committed by Indians.

The faithlessness and deception of the whites cannot be denied. The fact that whites intimidated and murdered Indians and stole their lands, and that they systematically and officially deprived Indians of their means of livelihood is equally true. On balance, it seems prudent to attempt to report the facts without assigning blame. Meanwhile, we can hope to learn from the past while we attempt to reconcile cultures that could not possibly understand each other in the past. As a side note, I offer the thought that capacity for understanding between individual Indian cultures, and between individual white cultures as well, has often been lacking. Perhaps we can be more accepting of each other's cultures in the future.

Plains Indian

©2002 by Timothy M. Leonard

I speak of love in tongues, in ancient dialects. Dialects of ancestors who lived where you are now for 12,000 years. Near the river all animal spirits welcome you with their love. They are manifestations of your being.

I am blessed to welcome you here. You have walked through tall-grass veldt of love to reach me.

My plains are narrow and smooth in places, rocky in others. I am the soil under your feet. I feel your weight, your balance — your weakness and your strength. I hear your heart beating as my ancestors pounded their ceremonial drums. I feel the surging force of your breath extend to the horizon. Wind accepts your breath.

I am everything you see, smell, taste, touch, and hear. I am the wildflowers; the moss and forests spread like dreams upon your outer landscape. I am your inner landscape. I see you stand silent in the forest hearing trees nudge each other. "Look," they say, "someone has returned."

I love the way you absorb a song of brown thrush collecting moss for a nest. I am the small brown bird saying hello. I am the sweet throated song you hear without listening. At night two owls sing their distant melody and their music fills your ears with mystery and love.

I am the warm spring sun on your face filtered through leaves of time. I am the spider's web dancing with diamond points of light. I am the rough fragile texture of bark you gently remove before connecting the edge of an axe with wood. You carry me through my forest; your flame creating heat of love. I am the taste of pitch on your lips, the odor of forest in your nostrils, filling your lungs. It is sweet.

I am the cold rain and wet snow and hot sun and four seasons. I am the yellow, purple, red, blue, and orange flowers out of brown earth.

I am an old dialect of Cheyenne and Arapaho tribes.

We shared our dialects with the French and Spanish explorers when they came here. We lived in harmony with Anglo-Dutch pirates in the 16th century.

We taught all of them our way, our culture and our language. We accepted their teachings, enduring their desires to take more from the land than they needed. We struggled to understand their motivation and spirit. We were sad to see their greed.

We were forced to move many times. We followed our elders, our guardian spirit from the eastern forests to the plains. Our spirit stood strong against adversity as we struggled to maintain our dignity and way of life.

I respect the spirit energies. I hear with my eyes and see with my ears. I understand your love of the guardian spirit power. I am the ancestor speaking 300 languages. Now only 150 languages remain. Language cannot be separated from who you are and where you live.

I say this so you will remember everything on the plains, along wild free rivers and inside forests. I took care of this place, and now your love has the responsibility.

Chapter Eleven

2002 Texas Archaeological Society Field School — a personal account

The Texas Archeological Society (TAS) 2002 Field School campground consisted of a tenting ground on private property, a kitchen located in a trailer, and a portable stage complete with a projection screen and sound system. Upon arrival, I could see this would be a well-organized affair. The arrangements and planning carried out by an experienced staff would weave a thread throughout a week that ended far too quickly.

The directions to the campground came in an email and were more than adequate. Upon arriving I registered, received a packet of information, and waited under a large oak tree for the newcomers orientation scheduled for the morning of arrival. I had an unexpected and pleasant surprise: The Texas Historical Commission state archaeologist, Pat Mercado-Allinger, took the microphone and delivered the orientation. She gave each of the three main groups specific information: The surveyors learned of their general schedule and what surveying would be like; the excavators were briefed on their sites, necessary tools, and techniques; the laboratory assistants were given specific information on their chosen area of occupation. Excavators and lab assistants were encouraged to swap around, in order to gain a broader understanding of the archaeological process and the reason for all the record keeping that would be required.

The TAS has held its Field School annually for forty-two years, and it has its traditions. There are members who've made it to every single field school since the first one. As I wandered around the area, I noticed an easy familiarity among many of the attendees, assuming them to be the "old-timers" and staff. I was greeted with smiles and friendly hellos on every hand. It was clear from the beginning of my tour that a genuinely open and easy-going attitude was part of the culture. The only amenity that seemed in short supply was shade for my tent.

I settled for the sparse dappling of a larger than average mesquite tree and began to set up my new blue cabin tent that would soon be known as the Taj (Taj Mahal), for its spacious room and screened front porch. The temperature was in the nineties and the sky was crystal blue; it seemed a portent of things to come. I broke my first sweat, struggling with poles and fabric that had never met one another before.

Over three hundred participants had registered, and most of them had arrived before me. For less than $150, I'd registered for the full week of excavation, seminars, meals, parties, and camaraderie. At such a low price, I was not surprised to see that many families—some quite large—were in attendance. Ages ranged from elementary school-aged to grizzled graybeards like myself. The TAS Field School is the premier event for avocational archaeologists in Texas. It is promoted through every local archaeological association around the state and by associations in other states as well. Additionally, several people came on the invitation and strong recommendation of veteran attendees. Some came

The author at ease under the "shade" of a mesquite tree. Photograph by Susan Morton, 2002.

from as far away as New Jersey, Colorado, and Montana. Others lived just down the road, in Austin or Waco, and went home from time to time for a bit of air conditioning and a hot bath. Using up the time until supper, I searched the camp for familiar faces and bumped into several people I knew. After greeting a few old friends and making several new acquaintances, I had my first meal of the Field School.

While packing for the week, I'd concentrated on the part of the list of supplies that had to do with camping, sleeping, and excavating. I'd given short shrift to a few items on the list for which I saw no obvious need. My mistake. On this end of the experience, the need for some of the overlooked items seems more than obvious. Lacking a tray on which to carry food from the mess line back to my chair, and having spent a week balancing paper plates, bowls, and cups on my knees, I now see the wisdom of bringing everything on the equipment list. A little faith would have been justified in this case. On the other hand, I developed a degree of coordination greater than I'd suspected was possible. Each meal offered two or three entrees, salads, fruits and/or vegetables, deserts, breads, multiple choices of beverage...a cornucopia of nutrition.

The facilities and arrangement of the camp gave evidence of both experience and forethought. The hand washing station consisted of a podium with liquid soap, and water provided in measured squirts with use of a foot pump. The men's and women's showers were positioned to provide maximum privacy—if not convenience. There were sufficient porta-potties strategically located and usually well serviced. RVs, tents, and pop-up campers were interspersed, with the exception made for those who wanted to run electrical generators: These were exiled down the road where the noise was less intrusive for the nonelectric campers.

During dinner I queried the experienced field school attendees whether I should set an alarm clock for wake-up and if so, what time I should get up. They informed me that I had no worries regarding the wake-up call: I would certainly hear it. The method of the wake-up call was left a mystery; no amount of prying would reveal the secret. The mystery was solved at 5:00 a.m. the following morning. Horn and stereo blaring, the Camp Boss drove a circuit that took him past every tent. Each morning was a different tune, ranging from the classical *La donna è mobile*, through the Beatles, and several (for me) unidentifiable country artists singing comic novelty songs—each one more or less appropriate to the events of the day. I had no trouble getting up on time, but the song of the morning did tend to stay in my mind all day.

Each day followed the same pattern. The crew chiefs assembled their crews and transportation was assigned. The two sites, one an established excavation project under the

Andrew Malof, principal investigator, inspects an expertly exposed feature. Photograph by Robert Marcom, 2002.

supervision of Andy Malof, known by its then-mysterious designation as "the snail site," and the other, a testing project under the supervision of archaeologist Elton Prewitt, lay only a few miles apart. Both are within miles of the Gault Clovis site. Educational and experiential opportunities were abundant. The pace was leisurely, by any standard, and the days were short. Projectile points, fragments of pottery and bones were common. A bison tooth and chert debitage (flakes and waste flint from tool-making) helped keep each excavator focused by offering evidence of past human activities. Nothing focused the attendees as effectively as the projectile points, though, and many were found.

The points were common—although no amount would have been sufficient, each one inspiring a lust for more—and ranged from a typically assigned age of several hundred to several thousand years old. The earliest type of point found by the testing crew was a Gower, which may date as early as 6,000 to 8,000 years before present (dates from *A Field Guide to Stone Artifacts of Texas Indians*, 2nd Ed., by S. Turner & T. R. Hester, Gulf Publishing Co., Houston, TX—an excellent resource for study of these and other Texas projectile points). There was ample evidence for campfires at both sites. Fire-cracked rock was abundant; charcoal and evidence of color changes on limestone rocks due to intense heat was seen in features identified as hearths.

In addition to excavation, TAS field school participants had a chance to find and record new archeological sites through survey. Survey participants systematically walked in teams across chosen landscapes hunting for the material remains of past human activities, such as burned rocks and chert flakes. Survey teams learned how to recognize sites and thoroughly document them through completing sketch maps, photographs, and site forms detailing the environmental setting, kinds of artifacts present, and their distribution.

While the grownups were digging, surveying, or participating in the lab, the younger archaeologists had the opportunity to excavate in search of a 150-year-old lean-to, a structure that may have been used by the original builder of the log house (now occupied by the landowner, Dr. Michael Collins) on the campsite property. A staff of archaeologists, both professional and avocational, oversaw the children's excavation, providing instruction and rewards in what I can only describe as a patient and inspired manner.

A feature identified as a hearth. J. L. Michael Williams, 2002.

The log cabin residence at the TAS 2002 Field School campground. Photograph by Robert Marcom, 2002.

Among the many traditional activities offered at camp, at least three must be noted: The fine seminars and presentations, the Wally Party, and the Friday Night Awards. The nightly presentations spanned the gamut, from flint knapping and fire-making exhibitions to mammoth excavations in the city limits of Waco, Texas; from Indian pottery techniques to bow making and atlatl throwing. Every afternoon and evening presented a unique opportunity to learn from very qualified professional and avocational archaeologists. These would have been sufficient to justify the price of attendance alone, and as they say on the TV infomercials: But wait—there's more!

No TAS Field School experience is complete without the Wally Party. The tradition was originated informally by a field school participant, now deceased, from whom the party takes its name. Beginning as a simple function offering rather potent margaritas (a drink composed of tequila, triple sec, and Mexican limes) and conversation, the event has grown to become a full-fledged party with live music and impromptu dancing exhibitions. The Wallyville sign on a white trailer informs partygoers they have arrived at a suburb of Margaritaville. The drinks are offered free to those who cannot afford the $1 per drink donation, and they are not as strong as they once were. They are very tasty, and the salty drinks will definitely hit the spot after a week of heavy perspiring.

The week wrapped up at the Friday Night Awards ceremony, where attendees are honored for such feats as remaining amazingly clean during the field school, traveling the furthest distance to attend, and for being the most creative in informal interpretation of artifacts, such as the discovery of a snail cocktail fork. Sometimes hard physical and intellectual work can be its own reward, but it can be gratifying to be rewarded by one's peers. One of the awards offered at the ceremony was in a serious vein and completely deserved: May and Jim Schmidt, two veteran avocational archaeologists, each received a plaque honoring more than 1,700 combined hours of service at the Gault site.

The two principal investigators, Elton Prewitt and Andy Malof, offered a summary of the weeks' excavations, and the more spectacular artifacts were offered for viewing. While interpretation awaits the conclusion of lab work, the list of projectile point types suffices to give some idea of the accomplishments. The points were found in a number of different types of site, including rock shelters and open and buried hearths, and were both solitary finds and those found associated with other artifacts. The types included Scallorn, Perdiz, Alba, Frio, Fairland, Pedernales, Ensor, Castroville, Montell, Gower, and several undefined arrow and dart points.

Several bags of artifacts are ready to be cataloged at the Field School lab. Photograph by Robert Marcom, 2002.

This tray contains several "unique" artifacts: a projectile base; a complete projectile point, and a drilled or pierced shell. Photograph by Robert Marcom, 2002.

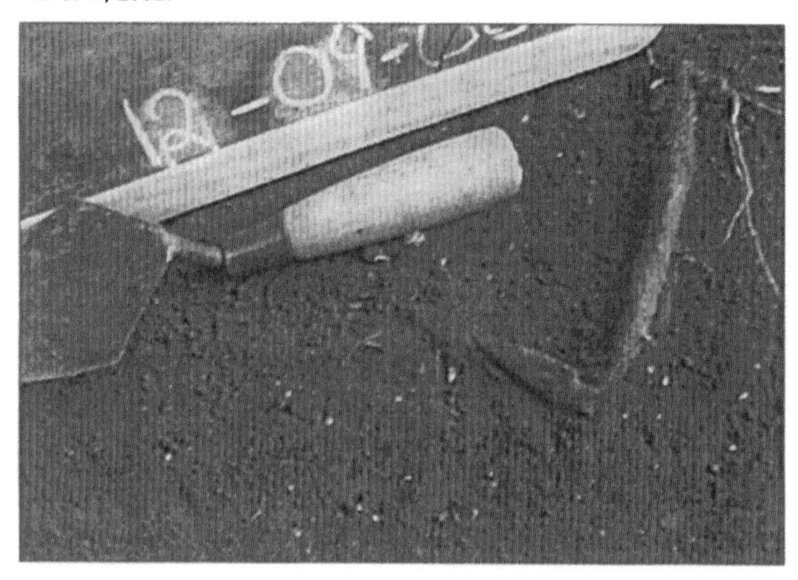

A Darl dart point. J. L. Michael Williams, 2002.

Ensor dart point. Andrew Malof, 2002.

Fairland dart point. Andrew Malof, 2002.

Montell dart point. J. L. Michael Williams, 2002.

Perhaps my personal highlight during the Friday night ceremony came just after the "graduation" portion of the program. Those attending TAS Field School for their first time were asked to come to the front and receive the "I survived 2002 Field School, Arkyologist" mantle, comprised of black permanent marker on orange flagging tape. Although I was a first-timer, I decided to remain seated, since there appeared to be plenty of graduates in attendance. Giving evidence of their great attention to detail, my absence was duly noted, and after the official ceremony, I was awarded several mantles.

The TAS Field School is a family experience that I can heartily recommend. The week was packed with new information, education, and acquaintances. The scheduled events even included a star party by the local astronomy association. Short workdays, extracurricular events, and late-evening conversations combined to create a truly memorable field school. The TAS website: http://www.txarch.org/

will offer information and updates on future field schools and other activities. TAS membership is required.

Interview with Margaret Howard, president of the Texas Archeological Society

Q: Can you tell us a bit about the philosophy behind the TAS Field Schools?

A: Professional and avocational archeologists collaborate at TAS Field Schools to provide hands-on training in good archeological practices for persons from the ages of 7 to 97. While education is the primary goal, recovery of significant information about the past is another important aim. Many field schools are held at sites threatened by natural or human processes like erosion or vandalism, and/or where no public funds are available to support excavations. Because of this focus, field school participants help to recover important information that would have otherwise been lost.

Q: What types of sites have been investigated by past field schools?

A: Over the past 40 years TAS Field Schools have been held in every region of Texas and at almost every kind of site imaginable, including coastal shell middens, missions and presidios, agricultural villages, rock shelters, rock art, plantations and sugar mills, prehistoric pueblos, historic Indian camps, rock-lined earth ovens, and historic house sites. The age of these sites ranges from 10,000 years ago to the 1800s. The incredible variety of field school sites offers TAS members—professional and avocational alike—an opportunity to polish their skills and knowledge of Texas archeology.

Q: What other activities does TAS offer?

A: TAS holds an annual meeting every fall at various locations around the state, organized in cooperation with the many regional archeological societies. The annual meeting includes scholarly papers by professional and avocational archeologists, workshops, a public lecture, and other opportunities to share information with persons interested in Texas archeology. New training opportunities will soon be offered, supported by the educational initiative of the 2002 Strategic Plan. These may include spring regional conferences and training sessions on topics like artifact analysis and report writing.

Q: Do you see avocational archaeology as an important adjunct to professional archaeology?

A: We are very fortunate in Texas, because these two groups often work hand in hand to protect the archeological resources of our state. Professionals offer skills and broad-based technical knowledge, while avocational archeologists are intimately familiar with particular areas. Also, many young people join TAS as avocational archeologists and have their first archeological experience at a TAS Field School, later going on to obtain college degrees and become professional archeologists. Professional archeologists in Texas soon learn that when they work in an unfamiliar region, their first priority should be to seek advice from local avocational archeologists.

Q: In your opinion, what are the most important aspects of the service avocational archaeologists perform?

A: Avocational archeologists outnumber professionals by at least 3 to 1 and are perfectly positioned to reach out to citizens and organizations in their communities. They can establish relationships with private landowners, allowing them to observe and record archeology in areas professional archeologists may never see. Many avocationals conduct scholarly investigations on private land where archeology is not required by state or federal law. Others carry out grass roots education and provide broad-based support for legislation to preserve archeology. Because almost all of the land in Texas is held privately, the role of avocational archeologists is very important.

Q: Would you recommend archaeology as a career?

A: Archeology is a great career for persons who enjoy the outdoors, who are curious and observant, and who are willing to learn. Because experience is the best teacher and Texas is so diverse, it takes years for a person to become an expert in a given area, and even more years to do competent archeology across the state. Even today, a number of areas and topics have not been fully explored, and new discoveries are still being made. TAS offers one of the many field schools in our state where people can obtain training under mentors who are willing to share their accumulated knowledge of archeology.

Q: Is there anything else you'd like to offer the reader regarding TAS or archaeology in general?

A: TAS offers a great opportunity for people to discover the many fascinating stories behind the objects left by early Texans. TAS can give every person who is willing to learn the skills needed to investigate life in ancient Texas and the ability to create a record that will live on to benefit future Texans.

Chapter Twelve

Public Archaeology

The point of contact between the academic community and the general public is one of the most important areas in the field of archaeology. While it is often perceived as a dividing line by both the professional archaeologist and by the lay public, it appears very different from each perspective. The perspective of the public is heavily colored by a sense of adventure, carefully cultivated by Hollywood to promote such products as "Laura Croft, Tomb Raider" and the Indiana Jones movies and TV series. The professional archaeologist, on the other hand, will tend to see excavation as a necessary stage to a largely structured exercise.

The old inclination of academic archaeologists to be wary of general interest was responsible for the top-down, didactic attempts to "educate" the public. Public archaeology was seen as a way to protect sites and artifacts by informing the laypeople of the importance of site integrity, thus discouraging pillaging. While the "public education" approach certainly had some success, it is also responsible for a certain degree of antagonism. The average person does not like to be told he or she should simply leave alone that which he/she does not understand.

Still, there is a legitimate concern on the part of professional archaeologists. Many valuable sites have been looted for their treasures over the years. Pot hunters have destroyed enormous amounts of information by removing artifacts. When the relationship to other artifacts is lost, knowledge is lost. The professional archaeologist can often recount one horror story after another, detailing sites robbed and priceless human heritage permanently erased. The federal and state governments have moved to protect the remaining historical and archaeological heritage within the United States through Cultural Resources Management, or CRM. This system of statutes and agencies goes far to protect the cultural information discovered at construction sites and on public lands.

The Texas Historical Commission Stewardship program attempts to extend CRM to the level of private land ownership. While historical and archaeological materials on private property cannot be protected by mandate in Texas, the THC offers extensive help and support for landowners. The THC archaeology stewards have as their mission to "preserve and interpret the vast archaeological landscape of Texas—covering 266,807 square miles and 254 counties." THC invites nomination (self or other) of candidates for stewardship.

Stewards are assigned to one of the six archaeological areas into which Texas has been divided: Mountain, Plains, Forts/Hill Country, Lakes/Brazos, Forest, Independence/Tropical. According to the information provided on the THC website, (http://www.thc.state.tx.us/stewards/stwdefault.html), "One of the most innovative and successful programs of its kind, the Texas Archeological Stewardship Network has served as the model for similar programs in other states. Stewards are not professional archaeologists but highly trained and motivated avocational archaeologists who work on a strictly volunteer basis."

The THC offers landowners assistance with obtaining National Registry designation and markers. THC offers the option of anonymity should the owner wish it. During the course of writing this book, consultation with archaeologists at all levels has rung with their consistent plea that sufficient credit be given to those who own the sites and allow exploration, excavation, and interpretation to occur. More information may be obtained from the Texas Historical Commission, or by visiting the website: http://www.thc.state.tx.us/markersdesigs/madsite.html.

Archaeological Societies

The area called public archaeology has come to encompass much more than CRM. Hundreds of Texans have joined regional archaeological associations, and there is room for hundreds more. Local associations often conduct excavations and labs under the supervision of state-licensed archaeologists. The membership attends regular meetings where the association may host archaeologists speaking on topics that range the archaeological gamut, from classical digs in far away lands to interpretation of patterns in recently discarded refuse. Local associations often provide the volunteers for ambitious projects such as the Red River War Battle Sites project. The activities of regional associations encourage and support many who aspire to avocational or professional archaeology.

Museums

Museum interpretation of archaeological excavations is another emphasis in public archaeology. They range in complexity and ambition from individual exhibits in big city museums, to those institutions dedicated to historical events or single locales, and to those that attempt a comprehensive interpretation for their area.

Museum interpretation has usually been intended to display particular artifacts in a visually satisfying context. The accompanying information is often sparse, depending on the creative artistry of a tableaux or diorama to convey the importance and context of the items on display. Many museums are now offering more in-depth interpretations through the use of free or low-priced publications.

A few museums visited in the course of writing this book are showcased here. This list is not intended to be comprehensive but rather representative. Many exceptional museums in Texas are not included. In the museums visited, the author found extensive resources helpful in the writing of this book

Among the best of those that interpret for a single locale is the **White Deer Land Museum** in Pampa, Texas. About the museum, Anne Davidson, curator, writes: "Exhibits include the authentically restored office of the White Deer Land Company, one of the Southwest's largest arrowhead and primitive tool collections, extensive photo archives including Native Americans, early ranches and settlements, and the first oil fields...." Admission is free, and donations are requested. The White Deer Land Museum is located in Pampa, Texas at 112-116 S. Cuyler. Its address: P.O. Box 1556, Pampa, TX 79066. Its phone number is (806) 669-8041.

Also in the Texas Panhandle, the **Panhandle-Plains Historical Museum** is located in Canyon, Texas, on the campus of the West Texas A&M University. The museum is "five museums in one with entire sections devoted to petroleum, western heritage, paleontology, transportation, and art." It is located 15 minutes south of Amarillo, Texas, at 2401 Fourth Ave., Canyon, Texas 79016. Its phone number is (806) 651-2244. Admission charges vary from free for 3 years of age and under, to $4.00 for adults and children 13 and over.

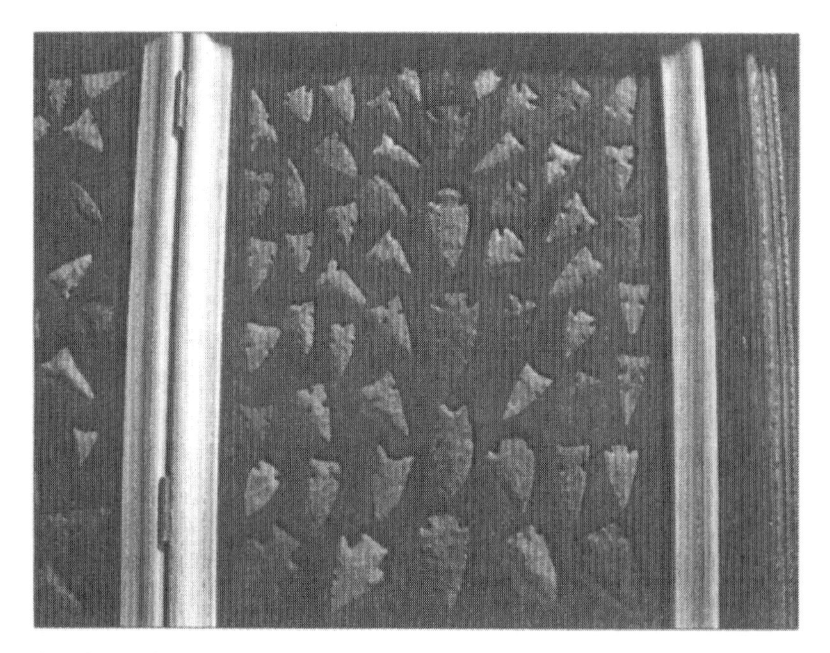

One frame from the extensive collection of projectile points at the White Deer Land Museum. Photograph by Robert Marcom, 2002. Courtesy of White Deer Land Museum, Pampa, Texas.

Museum of the Llano Estacado is located in Plainview, Texas, on the campus of the Wayland Baptist University. From the brochure: "The Museum of the Llano Estacado is a regional museum emphasizing the development of the geographical area of the Llano Estacado. The museum maintains 80 exhibits, in chronological order, describing geological, prehistoric, cultural, and economic development of the region." It is located at 1900 W. 8th Street, Plainview, TX 79072. Its phone number is (806) 296-5521 ext. 495. Admission is free; donations are accepted.

The Houston Museum of Natural Science is one of the premier museums of natural science and history in the U.S. In addition to exhibits and interpretation, the museum's curator of anthropology, Dr. Dirk Van

Tuerenhout, was extremely helpful and courteous in responding to requests for information and an interview.

Texas Historical Parks

State historical parks are another expression of public archaeology. Park managers, rangers, and volunteers are available to interpret the artifactual remains of Indian cultures, both ancient and historic, the Spanish and French exploration and colonization, the frontier forts system, the Old West, the oil boom and bust, as well as many other events that have occurred in this vast state. These parks, following, provided valuable assistance to me for the purpose of researching this book and are an excellent resource for further research into the sites included here.

For more information on the individual parks, please see the **Texas State Parks website**: http://www.tpwd.state.tx.us/park/parks.htm.

Prehistoric Sites

Caddoan Mounds State Historical Park

The park is located 6 miles southwest of Alto, Texas, on State Highway 21. Three 1,200-year-old mounds are located in the park. The interpretive center includes display and interpretation of artifacts and replicas of artifacts found during excavation. **Caddoan Mounds State Historical Park** (SHP) is the site of an ongoing experimental archaeology project. **Mission Tejas State Historical Park** is just a few miles away, also on State Highway 21, and offers a reconstruction of the first Christian church to be built in Texas. See the Texas Archeological Research Laboratory (TARL) website for continuing developments at http://www.utexas.edu/research/tarl/research/current.html #discoveries.

Caprock Canyon State Park and Trailway

Located 100 miles southeast of Amarillo, Texas, this park offers camping, hiking, and interpretation for both historic and prehistoric events. The Lake Theo Bison Kill site is located in the park. The park entrance is 3½ miles north of State Highway 86 in Quitaque, Texas, on FM 1056. The park manager, Geoffrey Hulse, and his staff were exceedingly helpful in assisting me to gain an appreciation for the long history of the Caprock Canyon area and are eager to do the same for all their visitors.

Historic Sites

Presidio La Bahia

Administered by the Catholic Diocese of Victoria, the Presidio at Goliad is a private institution dedicated to the preservation of this National Historic Landmark. Constructed on its current site in 1749, the presidio played roles in the Spanish, Mexican, and Republic of Texas historical eras. The presidio's original location on Garcitas Creek was also the site of the French Fort Saint Louis. Presidio La Bahia offers one of the finest collections of artifacts and interpretive presentations in Texas. I am deeply indebted to the administration and staff for their cooperation. Located 1½ miles south of Goliad, Texas, on Highway 183, the presidio is open daily. For more information, phone: (361) 645-3752, or write Presidio La Bahia, P.O. Box 57, Goliad, TX 77963.

Levi Jordan Plantation (LJP)

The Levi Jordan Plantation Historical Society was formed for the purpose of preserving the archeological and historical remains of the plantation and for the purpose of interpreting the history of all those who lived and worked at the plantation. I had the distinct honor of excavating at LJP during two University of Houston Archaeology Field Schools and on other occasions as well. My experiences at LJP were formative, both in archaeological skills and understanding, and in developing an acute awareness that through the process of digging up artifacts, we are digging up past lives and events.

The LJP Historical Society has extensive plans for exhibits, recreations, and tours. If you would like to become involved, visit the website at: http://www.webarchaeology.com/html/howtoget.htm and fill out the on-line feedback form.

Current Trends in Public Archaeology

Understanding the current trends in public archaeology is a challenge. I am fortunate to have interviewed Carol McDavid, archaeologist and doctoral candidate at Cambridge University, England. Carol has been concerned with the field of public archaeology in both an academic setting and in its practical application. The following interview explores McDavid's experiences in the practical application of developing public support for (at that time) an ongoing project.

The effort to manage cultural resources and to present the historical facts, and to do so with the approval and assistance of the descendants of both landowners and slaves/freedmen, demonstrates the modern demands placed on those concerned with the field of public archaeology.

Interview with Carol McDavid:

I'm interviewing Carol McDavid, who is an archaeologist, and we're talking about the interviews that she did with descendants of people who were associated with the Levi Jordan plantation in Brazoria County, Texas.

Q: Carol, you undertook to interview people in order to collect oral histories, is that correct?

> **A:** No, not exactly. In 1992 I was recruited by Ken Brown [Dr. Kenneth Brown, University of Houston, Ed.] to work on the public aspect of his archaeological work, and it was apparent that we needed to start publicly interpreting it, that is, to start talking about it with members of the public. So what I first did was just try to find out about the political situation—that is, was it going to be possible, feasible, and accepted in the community to do a public interpretation of this archaeological project. The house is still standing, as you know. Would they accept talking about slavery in their community and post emancipation, which as you know, is the period of time mostly covered by this excavation? Would they be comfortable talking about it? Would they support it? And would we be able to get support from African American and European American descendants? And so my job then was really to find out about, in Mark Leone's terms, the "conflicts and tensions" [1987, "Towards a Critical Archaeology"] that existed in the community. I wanted to find out whether people would support such a project. And so that's why I started talking to people. And the original idea was to interview people, get their opinions about what should or should not happen, and then, in that process, to hopefully find some people who would be

interested in actually helping to make it happen—like people who would be willing to join the board of directors and what not.

I started with a couple of assumptions before I even really talked to people much. One of these assumptions was something I actually got from a woman named Suzanne Spencer-Wood, who has done a lot of work in feminist archaeology—that was the idea of interpreting the past in a "both/and" way rather than an "either/or" way. That is, most of the plantation tours you're familiar with, across the South, have been traditionally "white people" tours—frequently with a picture over the mantle of the son of the owner. They talk about how he went to Harvard, and so on...and then if you ask the names of the people who lived in the quarters or worked in the house, they have no idea. You know, a "hoop skirts and mint julep" kind of tour. And in the early nineties, when this work started, that was still the dominant way of doing it, although they had started doing some things differently at Williamsburg and at Stagville Center and a couple of other places. So, to find out about this personally, I did a tour across the South—the good, bad, and the ugly tour across the American South. I had all this great information about plantations elsewhere to talk to the people in Brazoria about. And so I started getting names, and for the most part, I was able to talk to white folks pretty easily. But I needed to get contacts in the black community. I talked to a couple of people, and I was referred to a woman in the black community. I asked for an appointment, and she responded, "Why do I want to talk to you?" I replied, "We want to be sure we have people's opinions." And then she said, "So many people come down here and do research and then it stays on the shelf; what advantage will this have for my community? Why should I give you any information at all?" She was very polite and gracious, but not interested. So, I thought, I'm approaching this the wrong way.

At that time we had a nonprofit corporation, with a three-person board of directors. None was African American. My strategy originally was to interview people and get a sense from that which people in the African American community might want to serve on this board. But what I figured out, after this person said this to me, was that until we had members of the African American community on the board of directors—that is, empowered to direct my work and to have real power in running how this project was going to happen, I wouldn't be able to get much information from people. We had put the cart before the horse. We were trying to get people's opinions without them being empowered to have any control at all over what we wrote about them and what we did with the information. So, I talked to the other board members—two of the white descendants and the archaeologist in charge of the project—and said, "I think we should expand the board first before doing anything else." Really, the issue of power—empowerment—was the thing that started driving my work.

Q: You were involved with the dig at the Levi Jordan plantation from the early nineties. I know you and I worked there together in a field school, and you had some involvement there before I came out.

A: Not much though. I think that was my first full season.

Q: So, you not only worked on the public aspect of the project. You got your hands dirty—you dug. You saw the artifacts come out of the ground. You were around for the initial interpretations. You had this information when you went to talk to people about presenting the site to the public. Did that make a difference to you that you had been physically involved in the dig? Did it get in your way or help you to explain the importance?

A: It didn't get in my way. I probably could have explained the importance of the archaeology without having digging experience because the person who taught me about the site is very passionate, articulate, and persuasive: Dr. Ken Brown, as you know. So a lot of the information I had was secondhand until I really started excavating myself. And even at that point, Dr. Brown was really responsible for the primary interpretations. I think it is true, though, that the digging experience enhanced my feeling for the material. In that field season, you and I were excavating in a very interesting unit and found one of the objects that have become icons—the carved shell cameo. By the time I was interviewing people, I had been exposed to the first-person experience of finding remnants of the past in the ground. So, certainly, that tended to make me more passionate about it.

Q: Did your experience with talking to the descendants of the people who lived on that plantation, both in the quarters and in the great house, influence the way you looked at the dig when you went back to excavation? You met some of these people, you talked with them, they are real human beings and their antecedents who lived there must have become more real to you.

A: I think that's true with the European American community. The only reason it's not with the African American community is because—and I've tried to be clear about it in my writing about the project—the term descendant in this case really refers to the "descendant community" in a larger sense. It does not refer to the descendant community of this plantation. I got this insight from Cheryl Wright, who as you know got her master's degree at the University of Houston. By the time I started my work, she had already done her oral interviews of many of the people in the African American community, many of whom had ancestors who had lived on the plantation. She had to maintain privacy, of course, in terms of who her respondents were. One of the things that she found, and that I found later in talking to my contacts and other people, is that the idea of being a *site* descendant meant little or nothing to individuals in the African American community I met. It makes a lot of sense. Say your ancestors were enslaved on this plantation. Why are you going to carry that through history and be proud of it—in the sense of your affiliation with *that piece of land*? You wouldn't necessarily want to celebrate your family's affiliation with that particular plantation, even if you were proud of what they accomplished after they were no longer slaves. On the other hand, being identified as a member of a descendant community of a people who had been enslaved is different; it says more about how your ancestors survived that system. That is a very important distinction. For example, we have census data that indicate that one of our other very active board members is directly descended from a person who lived on the Jordan plantation. She does not necessarily accept that. She doesn't deny it, but she says her family never discussed it when she was growing up. My job is not to convince her that it's true. It's just that her connection to that particular place is irrelevant to her. She's an African American woman who is descended from people who lived in the community during slave times. She knows that. Her particular connection to the *Jordan* place is not something she grew up knowing. The point is not to convince people our information is right and theirs is wrong. The point is to start a dialogue about the past and to use this dialogue to relate to each other better now.

With the European Americans, I knew who they were and there was a very clear connection. In terms of African American descendants, we had to relate to them on their terms—in terms of the descendant community at large.

If you look at the history of the black churches in the area, you will find that there are names on those original church rolls from all three of the larger local plantations. My hypothesis (unresearched) about this is that after slavery, there were people from each plantation who were communicating and merging and beginning to see themselves as part of a *larger* community. But their communities are affiliated with the things *they had created*, their churches, not the plantations their ancestors came from.

Q: What do you think the Levi Jordan site will contribute to the community at large in Brazoria County?

A: That's the six-million-dollar question. I think it will contribute a lot, as public interpretations are able to take place on the site. The excavation is now officially complete. Dr. Brown is still doing artifact analysis. Right before the sale of the property, which happened fairly recently, he completed all of his on-site excavation work. The actual excavation lasted from 1986 into 2002. The only public interpretation of the site right now is the website and site tours that we do, and those don't happen very often. Once we're able to open it up to the public on a regular basis, I think the initial reaction will be extremely positive. I think people will relate to it, and I think people will find it empowering. I think it has great possibilities in terms of finding ways for people to look at the past in a different way.

Our original tours were attended mostly by white people, but as it has become more publicly known and understood that this project has been directed by a diverse group of blacks and whites, we have had more interest from black people in coming to look at the site. I think that very interaction, the idea of having a mixed group walking back to the quarters and talking about slavery together, is an important social activity that can counteract the stereotypes we, sadly, continue to have about each other.

Q: Would you see that as a healing opportunity, for different understandings to be aired in terms of what occurred there and why?

A: In the terms of Cornel West, one of my favorite writers, the Jordan site can enable us to create different "paradigms of imagination"—that is, alternative stories about the past. People will be able to see a past they didn't know was really there. The plantation says a lot about the way the site was abandoned and about the oppression of the nineteenth century. Obviously bad things happened there, and those things are not going to be ignored. On the other hand, the site also speaks a lot about how people figured out ways to control their lives and empower themselves despite their circumstances. To me, that's the important message. I think the people I work with, the people on the plantation board of directors, believe those are the messages they want to communicate. They are not interested in using the site as a way to further divisions in our society. We need to understand the roots of racism, and that site can help us understand that.

Q: You brought up the Internet. You have two sites on the Internet dealing with public archaeology.

A: Actually, the one I use is www.webarchaeology.com.

Q: I've been to your .com site. You present the concept of public archaeology through your own experience in collecting information and performing public interpretation. Do you have a further goal for that site?

A: Other than completing my Ph.D.?

Q: You are a candidate and you have completed your dissertation and are now in the process of defending it.

A: Yes. Part of the problem is that doing research about a process impedes the process. It creates an artificial situation.

Q: You can't investigate something without interfering with it?

A: No, you can't. I have been writing it up since 1999. Because my examiners are going to have to see the website as I used it at the time, I have not been able to go in and make a lot of changes. This is really at odds with the way the Internet should be. The idea of the Internet is to be able to make changes quickly and to have a site that is somewhat fluid. Well, that website's been "fixed," like a book, other than a few minor changes and corrections we've made along the way.

Q: Do you intend for it to be more dynamic?

A: Yes, I do. And the changes are going to be more driven by the local community even than the original site, which did have a great deal of community input. Brazoria, Sweeney, and West Columbia are the three small towns around the plantation. Part of my research involved going into the local school system and doing computer workshops with children there. Even though it's a small town in Texas, it has a really nice computer lab. One of the ideas is to have the kids rewrite the text on the website for kids. Right now, it's not really geared toward younger children. We want to actually involve the kids in writing about history—recreating a part of the website using the information that is there. I'd also like to update the archaeological data. This will depend upon whether Dr. Brown is able to help me in updating the data.

Q: What do you think the importance is of the information that is coming out of the Levi Jordan dig in terms of contributing to the understanding of the plantation system of the Old South?

A: For me personally, it has shown more about the connections between people who were enslaved and Africa. When people came from Africa, they came as people with very rich cultural traditions. Even though they may not have had material goods with them, they used material goods here in ways that they were able to cope with their surroundings. To me, that's the importance of it.

Q: Is there anything else you'd like to say about public archaeology that you think the general public might find interesting?

A: This is where my definition of public archaeology is different from a lot of people you may talk to. A lot of people think of public archaeology as archaeological education, and indeed at the end of the 1980s when the archaeological education movement really got going, it was a response to how much looting was going on at Native American sites, in particular. The idea was, then, that if we let people know of the importance of archaeological information, people wouldn't loot it. The idea was really conservation/preservation driven, and "public archaeology" was really created in response to a crisis situation.

Prior to 1990 public archaeology was mostly defined as CRM archaeology, or cultural resource management archaeology. That is, archaeology is funded by the public purse or mandated by law. CRM archaeology is archaeology that is mandated by law before property that is owned or financed by the government can be developed. If you have an office building and need federal financing, you have to do archaeology on the site to be sure there are no significant remains. Then later, in the 1990s, the term started being used in a more open, wide-ranging sense. I think this "opening up" process started in 1990 with the passage of the Native Americans Graves and Repatriation Act (NAGPRA).

Q: That act said if it's Indian, it belongs to Indians. It's a very powerful piece of legislation. You know how the government writes rules. There came to be known the process of consultation, wherein archaeologists who are doing work on native lands are required by law to officially consult with tribes about what they're doing. So you have this idea on the part of the descendants that they should be empowered.

A: Yes, and that really did not start happening in African American archaeology until well after the fiasco in New York City in the mid-1990s. So, public archaeology now has evolved. A lot of people are coming to think of it as *any* kind of archaeology that deals with, researches, theorizes, enacts, or manages the interface between archaeology as closed discipline and archaeology as the public perceives it. It can mean education and collaboration and consultation with native groups. It can mean cultural resource archaeology because that's funded by the public purse. It can mean being a media person in archaeology. Or it can mean doing websites about archaeology or being involved in local community archaeology, and so on.

Q: The very way that archaeological results are interpreted to people can be everything from the placard in the museum that describes the artifact and how it was used all the way through to the regulation of sites.

A: In my mind, it can include archaeological ethics as well. Of course, all of these areas should be a concern of all archaeologists. But all archaeologists don't have the desires, skills, or personalities to deal with the public on a regular basis, and to write for the public, and to create archaeological stories for the public. What you're doing with your book is a public archaeology project. Keeping in mind that I did some of my training in Europe, one of the things I saw there is that public archaeology over there really did not mean CRM. It meant archaeological heritage management, for the most part, which we really don't talk much about in this country. But that's how people do public archaeology in Europe. Museum work would come under this category. Pam Wheat, at the Houston Museum of Natural Science, also does public archaeology. It's any place where archaeology and the public interest connect.

Chapter Thirteen

State Historical Commission and the Archaeology of Texas

The Texas Historical Commission, the state agency for historic preservation, houses the lead office for archaeology in the state of Texas. Dr. James Bruseth serves as the director of the Archeology Division. The statewide nonregulatory programs of the division are supervised by the state archeologist. The regulatory programs are overseen by Mark Denton, and Steve Hoyt is the state marine archeologist. Dr. Dan Julien, also part of the Archeology Division, is responsible for the Texas Historic Sites Atlas, a computerized database of archaeological and historical sites in Texas. In January 2002 the state archeologist is Patricia Mercado-Allinger. Ms. Mercado-Allinger is responsible for coordinating a state program of archeological activities that spans varied interests.

The Texas Historical Commission (THC) is charged with nothing less difficult than preserving the human record of a vast, disparate, and complicated land. Texas is a land, complete with timeless identity and a strong "flavor," that is a powerful universal myth. The THC must manage the architecture, artifacts, documentary evidence, and archaeology of Texas. How can an agency of state government measure up to such a challenge?

With regard to archaeological resources, the Archeology Division of the THC exhibits professional competence, creativity, and even brilliance. One may be forgiven for suggesting that the programs of Archeology Division present a unique and valuable model for governmental involvement in preserving the artifactual record of the past.

Three recent projects by the Archeology Division have yielded excellent results: the excavations of the seventeenth-century Fort Saint Louis, the French ship *Belle*, and the Red River War project. The recovery of the sunken wreck of the *Belle* by use of a cofferdam was particularly impressive. This excavation required an engineering feat of great proportion.

The Archeology Division coordinated and supervised this very complicated project and the nearby excavation of the fort. The ruins of Fort Saint Louis lay buried on private land, as do much of the archeological remains in Texas. Permission was obtained for scientific recovery of this part of Texas heritage, not through the power of law, but rather by appeal to sensibility and sensitivity to private needs. The fort is located on a working ranch, and the excavations took place without undue impact on the ranch operations.

The third project mentioned above, the Red River War project, is nearly a polar opposite in both its nature and in location. Located on the high plains of the Texas Panhandle, this collection of battle sites contain artifactual evidence of the last of the struggles between the native people of the southern plains and the federal government determined to relocate them to reservation lands. Beginning with the battle at Adobe Walls in June of 1874, the Southern Plains tribes of the Comanche, Kiowa, Cheyenne, and Arapaho peoples sought to reestablish their control over the region. In a series of battles, they fought against the federal troops of the United States. By June 1875, the last of the hostile

Comanche were forced onto government reserve by the soldiers of the United States Army.

As with the Fort Saint Louis and the *Belle* sites, much of the evidence of this struggle between these two incompatible cultures lay on privately owned land. Interests were accommodated, activities were coordinated, and the heritage of both peoples is being recovered. In the process, history is benefiting from new information as it is discovered.

These excavation projects have been covered in detail elsewhere in this book. The enormity of both undertakings serves to illustrate the difficulty and importance of the work with which the Archeology Division is charged. The text of an interview with Ms. Patricia Mercado-Allinger is offered below, within the context of the preceding discussion.

Interview with Patricia Mercado-Allinger, State Archeologist

January 18, 2002

Q: Ms. Mercado-Allinger, can you give me a little bit of information about the variety of projects in which the Archeology Division engages?

A: Well, we have several programs. Special projects generally fall under the State Archeology Program—such undertakings as the current project in Victoria County, the Fort St. Louis excavations. Related to that project was the recovery of the *Belle* shipwreck in Matagorda Bay. Another current activity is the Red River War battle sites project that is focused on finding the locations of some of the key battle sites of the military and southern plains tribes that occurred in 1874 in the Texas Panhandle.

Q: Now when archeologists look at a battle site for instance, are they looking at it from the perspective of a historical event?

A: Certainly. We are very much interested in finding the material remains that are at the site and then comparing the data to the existing historical accounts. This evidence may support or perhaps refute what is stated in the documents. We're finding a little of both as we proceed with the Red River War project.

Q: So the written history is not necessarily factually accurate. It may be representative of cultural biases, then?

A: Remember, history tends to be written by the conquerors, not the conquered. So we're finding some discrepancies, and it's been an interesting and fascinating journey for us.

Q: The supervision of all archeology done in the state is done by your division, is it not?

A: The Archeology Division has some oversight over much of the archeological work that is done in Texas. I mentioned previously we have several programs. One of the programs involves state and federal review of projects that are to be undertaken on public land—federal, state, county, and municipal properties. Information about proposed development comes to the Archeology Division for review and recommendations by our archeological reviewers. This information could relate to construction projects of various kinds. Our reviewers consider the impact to archeological sites that are known or the potential for archeological sites in the affected areas. That's an important facet of the Archeology Division's responsibilities. Investigations may then be conducted by archeological contracting firms, with guidance from our staff reviews.

Q: Would it be accurate to say that the Historical Commission is a sort of clearinghouse for various interests—business interests, commercial interests, academic interests, perhaps federal and state interests?

A: Well there are certainly many things that converge here because development projects on public lands will oftentimes adversely impact archeological resources. Those are important considerations for the review. The Archeology Division also receives requests for information about Texas archeology from students of all ages. We try our best to respond to inquiries received from the interested public. We do a lot of things here in the Archeology Division.

Q: What about the idea of rescuing archeology sites that may be inappropriately used or excavated? Does the state have the authority to do that?

A: Well, in the state of Texas, the only laws that relate to the protection of sites are both the federal historic preservation laws and the Antiquities Code of Texas. The Antiquities Code really only relates to publicly owned properties of the state or any subdivision of the state. That would cover properties of the state, counties, cities, etc. Archeological sites on private property do not have any state law protections or restrictions. However, a private property owner can acquire legal protection for sites, because there are provisions for that in the Antiquities Code as well as with conservation easements. But there is no protection for sites on private property without written landowner consent.

Q: I see. So there's no statute that allows private property to be appropriated in any way, but the state invites private landowners to consult them, perhaps?

A: The Texas Historical Commission has no eminent domain authority. Instead, we work with landowners who invite us to assist them with archeological matters. This work is greatly facilitated by a special group of volunteers. The Texas Archeological Stewardship Network is an important volunteer program of the Archeology Division. The group is composed of about 100 dedicated men and women across the state. These "stewards" are avocational archeologists—they don't do this for a living. Stewards volunteer their time and expertise to provide local assistance, including emergency assistance, when we can't field a staff member right away. It has helped us, and I think in the bigger picture it has helped Texas archeology tremendously to have these men and women assisting us to preserve the archeological heritage of the state.

Q: It sounds as though the Texas Historical Commission and the Archeology Division in particular is interested in working with anyone to preserve archeological heritage in the state of Texas.

A: That's absolutely right! Yes.

Q: I'd like to ask about current or ongoing projects or near future projects for the state.

A: Well, certainly, the projects that I mentioned—Fort St. Louis, the *Belle*, and Red River War projects—are keeping us busy. When we leave the field, we really have only concluded the first phase of work. The projects that I just named are going to continue to draw a fair amount of our attention and energies. So much more work happens behind the scenes with the lab work, with the analysis of the data, and then of course the publication of results in various venues, technical reports as well as articles for the layperson.

Q: Is Fort St. Louis one of the earliest European military establishments in the state?

A: Certainly, yes. I believe that would be accurate.

Q: And the *Belle* was the ship that sank.

A: Yes, some of the survivors of the *Belle* then helped to establish the fort.

Q: And the Red River War would be referring to the battles between the federal army of the United States, the Texas militias, and the Plains Indians of the mid-1870s?

A: It was actually a campaign of the U.S. Army. Various columns converged on the Texas Panhandle from military installations in Texas, New Mexico, Oklahoma, and even Kansas. We don't really see any evidence for militia involvement during this campaign. The army engaged in various battles against such tribes as the Comanche, the Kiowa, the Southern Cheyenne, and Arapaho at various locales across the Panhandle.

Q: That covers quite a bit of area, doesn't it?

A: Certainly, although the battles seem to be concentrated along the eastern half of the Panhandle, from the headwaters of the Red River up to the far northeastern quadrant of the Panhandle.

Q: Could you explain a little to me of what the state archeologist does? What is your job?

A: In 1965 the Texas Legislature established the State Archeology program, and by virtue of that, a State Archeologist position. The program was originally placed in what used to be known as the State Building Commission and eventually moved into what is now the Texas Historical Commission. The enabling statute was very broad in the way it explained what this program was to accomplish. It essentially said, "establish a statewide archeology program" to include such things as site inventory, preservation, investigation, and coordination with allied groups. So it was very broad. Curtis Tunnell, the first state archeologist, concentrated on investigations of significant sites. That's when the first scientific excavations were conducted at some of the Spanish missions in San Antonio. Excavations at the Alamo were undertaken during his tenure.

Q: Now, the archeology of the state of Texas is a widely varied collection of subjects.

A: Yes, that's actually an understatement! To really help the public to grasp what we have in the state, we recently focused some of our energies on public outreach. With the support of the Council of Texas Archeologists, which is the professional organization, and the Texas Archeological Society, we coordinate an annual observance, Texas Archeology Awareness Month (TAAM). TAAM is observed every October, when there's a real effort across the state by various groups and institutions to hold special events for the public to heighten "archeology awareness," to help Texans to understand that we have quite a rich heritage in Texas.

Q: So, is public education high among your goals?

A: Absolutely. Because frankly, it's hard to preserve something when people don't know it's even here and that it's worthy of attention. Public awareness is also critical for continued support of archeological programs.

Q: The archeology of the state of Texas covers at least 12,000 years of human activity here. Does the department have any particular emphasis or does it work in all areas as the opportunities present themselves?

A: We seem to be focused on historical sites research right now, but as in the past, we certainly have taken advantage of opportunities to investigate Paleoindian sites and other prehistoric sites. Our work depends on the circumstances, challenges, and the opportunities that present themselves.

Q: Do we have any unique archeology in the state, either prehistoric or historic?

A: Oh yes, absolutely. Because we are such a large state, we encompass different environments. Cultural groups developed differing adaptations to specific environmental conditions through time, and that's reflected in the archeological record. So we definitely have tremendously diverse archeological resources in the state—everything from Southwestern archeology in far West Texas to Mississippian-related Caddoan mounds found in East Texas. And it runs the gamut.

Q: And then, of course, climates have changed over time as well.

A: That's correct.

Q: Going from an ice age, the end of the Wisconsin Ice Age.

A: Yes.

Q: Would cultures have had to adapt as some fauna disappeared and as others evolve?

A: There certainly is evidence in the archeological record for changes in the availability of animal and plant resources over time, and climate had a lot to do with that. There are some sites in West Texas that contain prehistoric well features that were dug during the periods following the Pleistocene Ice Age, when arid conditions prevailed. That is a single example of how archeological investigations can shed light on how prehistoric people were trying to find the necessary resources for survival.

Q: So, would it be fair to say that the Archeology Division has really been key to preserving evidence of cultures that could be obtained no other way than through archeology?

A: Yes, but keep in mind that there are anthropology departments in the state's various academic institutions as well as contracting firms that undertake archeological projects. In addition to the research projects that we direct, we are also involved with many archeological investigations that are undertaken in the state as a result of the regulatory reviews that are accomplished by the Archeology Division review staff.

Q: Are there any future plans that you could discuss with us about activities for the Archeology Division?

A: Well, we need to finish the ongoing projects. I can also tell you that the stewardship network is an important focal point for the State Archeology Program. We hope to continue to grow that program by expanding membership and providing the enrichment training that the stewards request. This training is designed to enable them to do a better job in their areas.

Q: You work with educational programs. You also have some programs of your own, opportunities for observation and participation?

A: At times. The circumstances vary from project to project. There are times when we can invite the public to observe what we're doing in the field or in the lab, as in the case of the Fort St. Louis Public Archeology Laboratory.

Q: Would the website be a good way for them to stay in touch with that?

A: Absolutely, and I can tell you that the Texas Historical Commission website is undergoing a major redesign so it should be even easier to access information about the agency.

Q: If you could make a list of two or three of the top accomplishments of the Archeology Division, what would you put on that list?

A: I think making more people aware of their archeological heritage is a key goal for us. That can be achieved in various ways. I think every staff member works to achieve this goal in one way or another, whatever role they play, be it in the review of development projects or through reports on field research that we undertake.

Q: Isn't it true that the Fort Saint Louis Public Archeology Lab in Victoria is unique?

A: Yes, it was very unique. Sadly, it is now closed because that phase of the Fort St. Louis project is winding down, but it was a wonderful way to expose the interested public to a sneak peek behind the scenes—to see what happens in the lab. You could actually observe artifacts coming in from the field and being processed.

Q: And then of course, the Texas Commission and the Archeology Division were essential in funding certain excavations that might not have happened otherwise, were they not?

A: Well, seeing that work was done. We don't necessarily fund a lot. Referring specifically to the *Belle*, Fort Saint Louis, and Red River War projects, we had to raise substantial funds from private and public sources to support these projects.

Many contract projects that are conducted as a result of our division's reviews are funded by the project sponsors—either a public agency or the permittee.

Q: It's not necessarily state funding?

A: Not always.

Q: Do you coordinate activities?

A: On the basis of project reviews, staff reviewers make recommendations for work that needs to be done to other state agencies, to federal agencies, and the like. The resulting investigations can be very important projects that add substantially to our knowledge of the past. I guess that would be another key goal worthy of mention—adding to our knowledge of the past.

Q: Now your department publishes as well.

A: Yes, we do. We produce a biannual newsletter, and as projects are completed, technical reports are published. Staff members also submit articles to scholarly journals as well as more popular venues.

Q: Is there anything else that you think would be particularly interesting or useful for people to know about your endeavors as state archeologist, your division, or the State Commission?

A: I would just add that the Archeology Division staff doesn't do it all. It's through coordination with other organizations, institutions, and individuals that our knowledge about the past and the work of preserving Texas's archeological heritage is accomplished. We try to provide leadership and guidance, but it all really can only happen with cooperation and coordination.

The Texas Historical Commission is located in a cluster of buildings in the Capitol Complex in downtown Austin:

THC Headquarters: Carrington-Covert House, 1511 Colorado

Architecture Division and Archeology Division: Elrose Building, 108 W. 16th

History Programs Division: Luther Hall, 105 W. 16th

Community Heritage Development Division: Christianson-Leberman Building, 1304 Colorado

Marketing Communications Division: Hobby Building, 333 Guadalupe

THC Regional Office: Sam Rayburn House Museum, Bonham

Staff members are available to answer questions and provide preservation assistance. Contact them at:

P.O. Box 12276
Austin, TX 78711-2276
Phone: 512/463-6100
Fax: 512/475-4872
email: thc@thc.state.tx.us
The website URL is: http://www.thc.state.tx.us/
Current Projects: http://www.thc.state.tx.us/index.html

Chapter Fourteen

Resources for Avocational Archaeology

A quick note on amateur excavations: please don't! I have no intention of encouraging looting, pot hunting, trespassing, or any other covert acquisition of artifacts. Opinions vary among avocationalists regarding surface finds (artifacts exposed through erosion), but I'd like to encourage readers of this book to adopt a strict policy against collecting. Artifacts from the past can reveal knowledge of lost cultures and techniques when placed in the hands of experts. If you find an artifact on your own land, it belongs to you. I would encourage you to share it with a local affiliate of the TAS. If you find an artifact on public lands or on someone else's land and you keep it, that is theft. The loss is greater than the simple act of stealing: The artifact you keep could have a greater significance by indicating the presence of unknown activities and peoples.

The stacks at your local library are a great place to start researching the archaeology of your area. You should contact the archaeological association for your region as well; many of them are listed below. An updated list may be found on the Internet, at the website of the Texas Archeological Society.

Speaking of the World Wide Web, there is probably no better source for information, and that applies to the field of archaeology, emphatically. I've included links, valid at the time of this writing, to a number of different sites. Please note, websites come and go rapidly. One of the best aspects of the web is that it quickly reflects change, but one of the most irritating aspects is that sites you found yesterday may be gone today. The best way to locate resources on the web is to use a search engine. Two of my favorites are: http://www.google.com, and http://www.lycos.com.

Other Sites of Interest

Texas Historical Commission Archaeology Awareness website:
http://www.thc.state.tx.us/archeologyaware/aastarch.html

THC Marine Archaeology website:
http://www.thc.state.tx.us/archeologyaware/aamarine.html

Texas Beyond History
http://www.texasbeyondhistory.net/

Texas Archeological Society website:
http://www.drsearch.com/
Email: txarch@onr.com
Postal Address:
The Center for Archaeological Research
U.T.S.A, 6900 N. Loop 1604 West
San Antonio, Texas 78249-0658

Web Archaeology Online:
http://www.webarchaeology.com/html/

TAS Affiliated Regional Associations

Oldham County Archeological Society
P.O. Box 635
Vega, Texas 79092

Panhandle Archeological Society
P.O. Box 814
Amarillo, Texas 79105

Dawson County Archeological Society
c/o Brownie Roberts
1507 North 11th
Lamesa, Texas 79331

Gaines County Archeological Society
c/o Garland Moore
Box 1581
Denver City, Texas 79323

South Plains Archeological Society
c/o Sue Shore
233 Indiana #E102
Lubbock, Texas 79415

Tarrant County Archaeological Society
P.O. Box 24412
Fort Worth, Texas 76124-1412

Collin County Archeological Society
c/o Heard Natural Science Museum
One Nature Place
McKinney, Texas 75069

Dallas Archeological Society
P.O. Box 600077
Dallas, Texas 75360-0077

East Texas Archeological Society
P.O. Box 630128
Nacogdoches, Texas 75963-0128

NE Texas Archeological Society
P.O. Box 239
Marshall, Texas 75671

Brazosport Archeological Society
Brazosport Museum of Natural Science
400 College Drive
Lake Jackson, Texas 77566

Deep East Texas Archeological Society
c/o Patricia Forrest
Rt 1, Box 213B
Call, Texas 75933

Fort Bend Archeological Society
P.O. Drawer 460
Richmond, Texas 77406-0460

Houston Archeological Society
P.O. Box 6751
Houston, Texas 77265-6751

Bee County College Archaeological Society
3800 Charco Road
Beeville, Texas 78102

Coastal Bend Archaeological Society
c/o Larry Beaman
303 Rolling Acres Drive
Corpus Christi, Texas 78410

Rio Grande Valley Archeological Society
Harlingen Industrial Air Park
Harlingen, Texas 78550

Webb County Archaeological Society
c/o Rose Trevino
102 Mize Drive
Laredo, Texas 780045-1986

Central Texas Archeological Society
4229 Mitchell Road
Waco, Texas 76710.

Llano Uplift Archeological Society
P.O. Box 302
Kingsland, Texas 78639

Mid-Texas Archeological Society
c/o Laverne Drews
Rte 2, Box 38
San Saba, Texas 76877

Travis County Archeological Society
P.O. Box 9464
Austin, Texas 78766-9464
The Commons Building (room to be posted)
Pickle Research Campus/UT-Austin
10100 Burnet Road

Concho Valley Archeological Society
c/o Larry Riemenschneider
Rte 2, Box 55
Miles, Texas 76861

Iraan Archeological Society
P.O. Box 183
Iraan, Texas 79744-0183

Midland Archeological Society
P.O. Box 4224
Midland, Texas 79704

Big Bend Archaeological Society
P.O. Box 1
Big Bend National Park, Texas 79834

El Paso Archaeological Society
P.O. Box 4345
El Paso, Texas 79914-4345

Southern Texas Archaeological Association
P.O. Box 791032
San Antonio, Texas 78279-1031

Hill Country Archeological Association
P.O. Box 290393
Kerrville, Texas 78029-0393
sanerjr@ktc.com

Concluding Comments

If I've managed to pique your interest in archaeology, then my work is done. All that remains is for you to get involved. If your interest is avocational, then join one of the local societies listed in the previous chapter. If you're interested in a career as a professional archaeologist, then begin a dialogue with the university in your area.

If you choose the avocational track, you don't have to dig in order to participate. Most of the local associations will welcome you at whatever level of participation you choose. I've found associations eager to share the awareness of our archaeological resources in Texas through presentation of their own work, the work of other associations, and through speakers and presentations. Of course, they will be happy to have you dig and participate in washing, labeling, and cataloging artifacts in the laboratory. They will usually be equally happy if you simply show up for meetings.

If you are considering attending college for an archaeology-related major, I'd like to recommend that you investigate your local community or junior college for the first two years of required academics and core curriculum. You may be able to take some of your major courses at the same time, but even if that doesn't prove to be possible, there are distinct advantages to beginning your college work at a community college: Classes will often be smaller, and should you need it, more assistance should be available than is common at a four-year university.

For the college-trained archaeologist, there are two career tracks available: Contract or Cultural Resources Management, and the academic or instructional track. Opportunities in both of these tracks will be enhanced by field experience that can be gained through membership in an archaeological association. Membership in the Texas Archaeological Society and your regional association as well may give you a broad sense of current projects and a base of information from which you can decide your specific interests. You will be exposed to the jargon, to the underlying philosophies, and to the many and varied lab and excavation methods in a supportive and instructional manner.

If you are a landowner with possible or confirmed archaeological sites, I hope I've both imparted a sense of wonder at the cultural treasure that reposes in your trust and reassured you with the fact that making your property available for archaeological survey does not compromise your property rights. You may feel assured that the Texas Historical Commission and the various universities and colleges as well as private contract firms all must respect your wishes. All artifacts found on your land belong to you.

If you are an "arrowhead hunter" or collector of artifacts, let me assure you that I understand the urges well. My interest in archaeology awakened with the first "bird point" I found at a highway cut in South Texas. That point, a beautifully formed harrel, sparked a romance and a quest for knowledge that continues to this day. As I collected more projectile points, scrapers, and knives, I sought information about them. I read books written for collectors, talked with those more knowledgeable than myself, and read up on every topic I thought was related to the people who left behind the artifacts I now possessed.

Perhaps the best thing that happened to me as a collector was the loss of my entire collection during a move. When I unpacked and found the collection missing, I was devastated. It was gone, I didn't have a clue as to where it ended up, and I had no heart for starting another one. However, the interest in Native Americans and their culture and societies remained.

While attending my first semester at the University of Houston, I noted an introductory class for archaeology in the Anthropology Department. That class opened a new world for me. I changed my major, acquired an academic counselor in the department, and filed a new degree plan within the week. While my academic career has had remarkable ups and downs, I've never been sorry for taking that major course of study.

Since completing the major courses (but choosing not to complete the Bachelor of Science degree), I've worked and studied in the field of archaeology and devoted my time to the field as an avocationalist. Even though I consider myself a full-time writer at this time of my life (I've grown too old and fat to follow the life of an archaeology shovel-bum), I continue to pursue the avocation. May I suggest, to all the collectors who've read to this point, that possession of finely crafted artifacts is a pale shadow of the satisfaction available through contributing to a deeper understanding of the people who made them. Your collections are extremely valuable, and I encourage everyone who has more than a few pieces to have their collection documented and photographed. I encourage you to go one step further, and join your local association.

Bibliography

The following books and publications were consulted in the writing of this book:

Archaeology Theories, Methods and Practice by Colin Renfrew and Paul Bahn, published by Thames and Hudson Ltd., London, England, 1991

Ancient North America by Brian M. Fagan, published by Thames and Hudson Ltd., London, England, 1991

The Indians of Texas by W. W. Newcomb Jr., published by University of Texas Press, 1990

The Red River Wars Battle Sites Project, Phase 1 by J. B. Cruse, P. A. Mercado-Allinger, D. D. Scott, P. Folds, published by Archeology Division of the Texas Historical Commission, 2001

The Red River Wars Battle Sites Project, Phase 2 by J. B. Cruse, P. A. Mercado-Allinger, D. D. Scott, P. Folds, published by Archeology Division of the Texas Historical Commission, 2001

Bulletin of the Texas Archeological Society, Vol. 72/2001 published by the Texas Archeological Society, the following articles:

"Querechos and Teyas" by Douglas K. Boyd

"The Great Kingdom of Tejas" by Timothy K. Perttula

Studies in Archeology 30 (1992) *The Native History of the Caddo* published by the Texas Archeology Research Laboratory, the following articles:

"The George C. Davis Site: Glimpses Into Early Caddoan Symbolism and Ideology" by Dee Ann Story

"Skeletal Biology of the Prehistoric Caddo" by J. C. Rose, M. P. Hoffman, B. A. Burnett, A. M. Harmon, and J. E. Barnes

Archaeological Excavations of Antelope Creek Ruins and Alibates Ruins Panhandle Aspect 1938-1941, published by the Panhandle Archeological Society, 2000

Southeast Texas Archeology by Leland Patterson, Houston Archeological Society Report No. 12, 1996

A Field Guide to Stone Artifacts of the Texas Indians, 2nd Edition by E. S. Turner and T. R. Hester, published by Gulf Publishing Co., Houston, Texas, 1993

The Texas Rangers, A Century of Frontier Defense 2nd Edition by Walter Prescott Webb, published by University of Texas Press, 1965

Adobe Walls Wars by Bob Izzard, published by Tanglaire Press, Amarillo, Texas, 1993

Historical Archaeology Vol. 24 (1990), the following articles:

"Structural Continuity in an African-American Slave and Tenant Community" by K. L. Brown and D. C. Cooper

"The Archaeology of Racism and Ethnicity on Southern Plantations" by D. W. Babson.

Index